# Autonom(
# Opportuniti(
# and Disruptions

## Michael E. McGrath

### First Edition
### July 1, 2018

**Dedicated to:**

**My Amazing Family**

# Table of Contents

# Introduction

People frequently ask me why I decided to write a book about autonomous driving. My answer is that it will be the biggest thing in the remainder of my lifetime, possibly the biggest thing to happen in this century. When I started researching autonomous vehicles, I did so mostly out of curiosity. Why was Google investing in self-driving cars? I thought it was a stupid waste of money, but nevertheless, I was curious. Or so I thought before I started my research. As I got deep into my research, I began to realize the magnitude the advent of autonomous vehicles will have on the world. It will ultimately transform transportation and society as we currently know it.

I became excited about how significantly autonomous vehicles will change our everyday lifestyles even more than smartphones or streaming video, because transportation is so central to our lives. I found that the technologies enabling autonomous vehicles are fascinating, and more importantly, they are all feasible. When I analyzed autonomous ride services, I was blown away by its economics. Most new technologies have benefits, but I found the benefits of autonomous vehicles, to be overwhelming – perhaps more significant than anything so far.

As an expert in the strategies of technology companies, I was drawn to discover the strategies that companies are pursuing for autonomous vehicles. What I found was one of the most exciting strategic competitions ever. There will be big winners and losers, actually very big winners and very big losers. The strategies these companies follow will achieve significant financial benefits, or failures.

I found the disruptions unsettling. The massive loss of jobs, and the number companies and industries that will be injured or destroyed by the advent of autonomous vehicles, is frightening. I rationalized that this is pure math: new replaces old and benefits come at a cost. It will be a massive transformation that will take place no matter what, because the benefits are so overwhelming.

This book has primarily a strategic and economic focus, but also provides an introduction into autonomous vehicles, how they work, their underlying technologies, how they will change lifestyles, and the benefits they will bring. My objective is to provide an overall perspective of the advent of autonomous vehicles. For those intrigued by autonomous vehicles, it provides sufficient background and description to understand them. For those working in the transportation or technology industries, it provides insight into the strategies companies might follow.

While providing what I hope will be an enduring perspective, the autonomous vehicle industry is developing rapidly. Therefore, I expect I will need to update this book regularly with new editions. So be warned that some of the forecasts and predictions will be refined or change over time.

This book is about the future. So, it contains forecasts and predictions. To better understand the economics of autonomous vehicles, particularly autonomous ride services, I developed several economic and business models. These models are directional, meaning that they are directionally correct but not precise. They illustrate the incredible economics of autonomous vehicles, even with a range of assumptions for critical variables.

I also make forecasts on the timing and magnitude of the new markets created by autonomous vehicles. These are forecasts, and most certainly will prove to be incorrect. Nevertheless, they illustrate the extent of the change coming and how fast it could happen.

Finally, I go out on a limb and try to predict the strategies for major companies. None of these use any inside information. I did these based on my experience in strategy for technology companies. Some companies have given some indication of their strategy. Others have been secretive. In many cases, I predict my own opinion of what their strategies should be, based on the market opportunities, their strengths, and their weaknesses.

For all these reasons, I decided to title the book *Autonomous Vehicles: Opportunities, Strategies, and Disruptions*. I hope you enjoy it!


Michael E. McGrath
michael_e_mcgrath@msn.com

# About the Author

Michael E. McGrath is an experienced technology consultant, executive, and author. He is a founder of PRTM, one of the most successful global consulting firms on technology development and strategy. At PRTM he created the firm's consulting practice in product development and strategy with the well-known PACE process.

He is a former CEO or Chairman of three publicly-traded technology companies. He led the turnaround of software company i2 Technologies in 2005 and later served as chairman of Entrust. He currently serves on the board of directors of National Instruments.

In addition, he authored seven books on strategy and decision making. His best-selling book, *Product Strategy for High-Technology Companies*, introduced strategic concepts for product-platform strategy, core strategic vision, vectors of differentiation, and leveraged growth. His decision-making books include: *Business Decisions, Decide Better for a Better Life,* and *The Wit and Wisdom of Decision Making*. His books on product development include: *Product And Cycle-time Excellence* and *Next-Generation Product Development*.

# Executive Summary

Autonomous vehicles will change our fundamental lifestyles, and in doing so create what are perhaps the most significant opportunities of this century. The benefits are unprecedented. The challenges are sizable but not insurmountable. The strategies are exciting. The disruptions will be frightening. In this book, I explain why I believe these bold statements are true.

I expect that many people will challenge these statements, just as in 1899 when the Literary Digest declared "The horseless carriage...will never come into as common use as the bicycle."

Transportation is involved in almost everything in our everyday lives, and autonomous vehicles will completely transform transportation. People will commute to work autonomously, using their new-found time for work or leisure. Restaurants and others will provide free car service for their customers. People who currently can't drive will have new mobility. All types of travel will change. Even food will be delivered autonomously. The potential change in lifestyles created so much enthusiasm that I made it the first chapter.

Chapter 2 introduces autonomous vehicles. It defines practical categories for autonomous driving, including the definition of an important new category that is critical to the strategic perspective. While it will take time for autonomous vehicles to master everything, the point that most people miss is that autonomous vehicles only need to be, what I refer to as, *sufficiently autonomous. Sufficiently autonomous* is a category of autonomous driving that means an autonomous vehicle can take you from one specific point to another – such as from your home to the airport or the office or the restaurant. That will cover more than 90% of the trips needed. This chapter also discusses the SAE levels of autonomous driving and describes the functions of autonomous driving from basic to semi-autonomous to sufficiently autonomous to fully-autonomous.

The benefits of autonomous vehicles (AVs) are mind-boggling. They will avoid 90% of the accidents, eventually saving hundreds of thousands lives and preventing injuries to tens of millions of people. AVs will reduce the costs of transportation by as much as 70% and save families thousands of dollars annually. The average person will be able to do things other than driving, giving him/her nearly 300 hours of "found time" annually. That equates to more than 250 million hours of "new time" in the United States alone. AVs will provide mobility to millions of people who cannot drive, improving their lives for the better. As I describe in Chapter 3, the sheer magnitude of these benefits is unprecedented, and it makes the advent of autonomous vehicles inevitable.

One of the major conclusions in this book is the significance of what I call autonomous ride services (ARS). There are many different terms for this, such as mobility as a service, transportation as a service, robo-taxis, etc., but I prefer to use this term. ARS will be huge! This new industry will achieve $750 billion in revenue in the United States alone, possibly much more, by 2030. It will also be the first significant use of AVs because they only need to be sufficiently autonomous for ARS. In Chapter 4, I clarify why ARS will be the primary new market for AVs. Then I introduce new economic and business models to illustrate how ARS businesses will work. They will deploy fleets of autonomous vehicles to provide these ride services in each major metropolitan area. The exceptional economics of ARS enable extraordinary profitability at a price of only $1 per mile, maybe less. The underlying reasons for this profitability are the elimination of driver costs and significant increase in vehicle utilization. ARS will be so inexpensive and convenient that many people will abandon individual car ownership. Most people will be able to summon a car whenever they want, travel in luxury, and the cost will be much less than owning a car. I explain how the autonomous ride service industry will emerge, and the fascinating economic models behind it.

Autonomous vehicles include trucks, delivery vehicles, and buses. However, as I describe in Chapter 5, autonomy will benefit these segments in different ways. Autonomous trucking will initially benefit from enabling the truck driver to let the truck drive by itself while he or she is resting and taking the required breaks. Long-haul trucking can see up to a 40% improvement in delivery time and truck utilization from this alone. Food delivery will quickly become autonomous because the cost savings and efficiency improvements are significant. Package delivery, however, won't benefit as soon because, unlike food delivery, there isn't always someone there to take the package from the truck. Autonomous shuttle buses also will see early benefits since they have short predetermined routs, but buses used for general transit may not change, and in fact, may be displaced by ARS.

Autonomous Vehicles (AVs) are enabled by some exciting new technologies that are developing in parallel. Up until now, the necessary technologies have been theoretical or in their infancy. In Chapter 6, I try to provide a layman's view of these technologies. AVs can see what's around then using an array of sensors. While there are different types of sensors, such as cameras and radar, lidar is perhaps the most important new technology. It may also be a key technology for AVs because it is expensive and getting cost reductions is essential. There is also some technical debate about the need for lidar.

Computers provide the brain for AVs. The computers used in AVs are specialized. They are very powerful custom processors with high-speed graphics processing, The Nvidia Pegasus, for example, incorporates four high-speed processors explicitly designed for AVs. It processes a fantastic 320 trillion operations a second. Computer software provides the intelligence for AVs doing sensor fusion, analyzing everything that surrounds the vehicle, planning its path, and instructing its moves. Artificial intelligence is critical to the AV learning process both in the vehicle and in simulations.

The market opportunity for autonomous vehicles is enormous. Some estimates show it exceeding $1 trillion per year. This opportunity is not a single market; autonomous vehicles will create several large markets and multiple market segments. The first significant market to emerge will be autonomous ride services. In Chapter 7, I attempt to define and estimate the size of these new markets and predict the strategies that each major company will pursue.

The companies pursuing these new markets see autonomous vehicles as an exciting opportunity, and they are investing a massive amount of money developing them. I estimate that more than $100 billion already has been committed to develop these technologies and bring these vehicles to market, and much more will be invested in deployment as the market is created. These companies are not stupid or naïve – they are among the best-managed companies in the world. They see exciting new business opportunities in AVs. The sheer magnitude of these investments makes it inevitable that autonomous vehicles will be brought to market as quickly as possible.

The anticipated strategies of significant companies are fascinating. There is a lot at stake in capturing these large new markets. What strategies will succeed? Which companies will capture these new markets? Even though some companies have not declared their specific strategies, I describe the strategy that I anticipate each major company will employ. Many companies will jump into the ARS market since it is going be so large and will be the first new market.

Waymo (Google) is developing autonomous driving technology and will be one of the first to provide ARS, modifying Chrysler Pacifica minivans and Jaguar I-PACE vehicles with Waymo technology as its first-generation ARS autonomous vehicle. Uber's ARS strategy is similar. It will use its own technology deployed on Volvo minivans. Lyft will employ a very different strategy. It is creating an open ARS platform for AV manufactures to use for ARS. Car manufacturers, especially Ford, GM, and Mercedes, will try to enter this market, but they will have challenges that I describe. Apple is the wildcard in ARS. Although it has kept its intentions to itself, I expect that it will become a significant competitor in ARS.

Individually owned autonomous vehicles will also be an incredible new market, but for reasons that I explain, I expect that it will grow more slowly than autonomous ride services. Tesla has been the pioneer of this market, but some traditional car manufacturers such as Mercedes, BMW, GM, and Ford are close behind. AVs require new technologies, and the companies competing to provide these technologies also have some exciting strategies.

There are also some frightening disruptions from the advent of autonomous vehicles, which I describe in Chapter 8. The enormous benefits come at a cost. The emergence of autonomous ride services will decrease personal car ownership. Approximately 17 million new cars are sold annually in the United States with a market of almost $1 trillion. I forecast a continuous decline in this market, as autonomous ride services cannibalize it. Using a model that predicts the growth of autonomous ride services, each new ARS vehicle will displace 8-10 new cars sold since individually-owned cars are only used approximately 5% of the time, while autonomous ride services vehicles will be utilized 50% of the time. The switch to ARS could eventually reduce new car sales to less than 10 million vehicles, most of which will be autonomous.

Many auto-related industries also will be disrupted, including car dealers, gas stations, auto repair shops, car rental companies, non-autonomous ridesharing, taxis, public transportation, insurance companies, and many others. In the broader transportation industry, millions of jobs for truck drivers and delivery drivers will be eliminated. There will also be reductions in the need for emergency medical services and significantly lower revenue to cities and states for traffic violations. Collectively, the advent of autonomous vehicles, particularly ARS, will result in the loss of tens of millions of jobs in the United States alone. I expect that a disruption of this magnitude will cause some to question if autonomous vehicles are a good thing. In the end, though, it's difficult to argue that it's not worth the savings in lives, injuries, time, and economic benefits. I believe that society, at least in developed countries, will need to rethink the nature of work.

Government regulation and support will play an essential role in autonomous vehicles. However, as I discuss in Chapter 9, I don't expect regulation will be much of deterrence because the benefits are so enormous.

Given the immense magnitude of the autonomous vehicle transformation, its projected timing is of great interest. I capture this into five-year stages of adoption in the concluding chapter. We are currently in what I call *Stage 0 (2016-2020) Development and Testing.* AV development is fully underway with more than $100 billion invested or committed, and the leading companies are already into testing. By 2019 and 2020, Waymo, Uber, and others will complete testing sufficiently to begin rapidly deploying ARS in Stage 1.

*Stage 1 (2021-2025) Launch of ARS* will be the most exciting stage for AV adoption. As the stage title implies, the focus will be on ARS. There will be what is essentially a land rush by significant players to deploy ARS fleets to critical metropolitan markets during this stage. I estimate that as much as $200 billion will be invested to establish ARS during this stage. By the end of this stage (2025), almost a million ARS vehicles will be deployed generating revenue of more than $150 billion. During this stage, there will also be significant adoption of autonomous vehicles in the retail market and trucking and delivery. The disruptions previously discussed will start toward the end of this stage.

*Stage 2 (2026-2030)* will bring in broad acceptance of AVs. ARS will continue to grow with revenue reaching $750 billion, and more than 15% of the miles traveled will be by ARS. The automotive retail market will continue to shrink, and by the end of this stage, more than 75% of the new vehicles sold will be at least sufficiently autonomous, enabling them to drive on their own most of the time. It is also the stage where the disruptions of AVs will become apparent to everyone.

*Stage 3 (2031-2035) More Advanced AVs* and *Stage 4 (3036+) AVs Completely Displace Cars* will be the next two stages in the adoption of AVs. By then the horseless-carriage, a little more than a century after its introduction, will be gone into history.

# Chapter 1
# New Lifestyles

Autonomous vehicles offer an exciting future, although most people can't envision it yet. Perhaps the best way to understand how autonomous vehicles (AVs) will change the ways we go about our lives and the ways we work is to imagine the new lifestyle models that they will create. AVs will change the way we commute to work, giving us new-found time. They will change the way we get to and from restaurants, activities, places, and events. They will enable significant lifestyle improvements by providing mobility for those who cannot drive. They will change the ways food and packages are delivered. They will make ground travel less expensive and more convenient than air travel in some cases.

In addition to individually-owned AVs, autonomous ride services, essentially Uber without a driver, will change the profile of transportation. I discuss this in detail in Chapter 4, but I use some examples here.

New lifestyles with autonomous vehicles are exciting. In this first chapter, I will describe imaginary, but likely, autonomous vehicle lifestyle scenarios. I hope it sets the stage for this excitement.

## Commuting

Commuting to and from work is tiresome for most people. It's non-productive time, often considered as wasted. AVs will transform commuting as we know it today. Commuting will be more productive, and provide an incentive for more people to live further from where they work. It will also reduce reliance on car ownership and public transportation. Here are some examples.

*AVs will transform commuting as we know it today.*

### Individually-Owned AV Commuting

Many individuals will use their own AVs for commuting to and from work, enabling them to replace their commuting time with more productive work or leisure activities.

*Joan and her husband Mark use their AV primarily for their long commute to and from work. They decided to buy a more affordable home further outside of the city where they both worked and then purchased a new AV to use the additional commuting time more productively. They usually leave for work at 8 AM and use their 60-minute commute to have coffee and muffins and read the morning papers on their iPads. They purchased the mobile kitchen option in their AV that provides a small coffee maker and refrigerator. They usually drop Mark off at work first and then Joan last because her company offers free AV parking. They often leave work at 6 PM and watch the news on the way home, sometimes having a glass of wine on the way. They are pleased to be able to afford a much nicer home further from the city, while not having to waste more time commuting. They have the house of their dreams for the same they would have paid for a much smaller home closer to the city.*

### Shared AV Commuting

Shared AV commuting will replace carpooling as we know it, eliminating the need for someone to drive and provide a car, or alternating drivers.

*Anthony shares an AV commuting vehicle with three co-workers. Autonomous Commuting Corp. provides this car for a monthly fee. They each pay $10 per day ($200 per month) for the 30-mile commute. At the cost of $0.16 per mile each, this is much less than the previous carpooling and parking expenses. The spacious AV is designed primarily for commuting with comfortable individual seating and high-speed internet. The car reliably picks them up at the same time every day. They each have their routine in the morning and the evening commute. They read, watch videos or TV, drink their coffee, or sleep. One of them is using her time to learn a new language. If one of them works late, he or she merely requests an alternative ride from Autonomous Commuting. Autonomous Commuting uses the AV throughout the day to transport others, enabling it to provide low-cost commuting.*

### City Commuting

Many people who work in the city will use an autonomous ride service to get back and forth to work, instead of owning a car or using public transportation. They won't have a scheduled pick up time or monthly contract. They will use an autonomous ride service (ARS) like the way people use Uber ridesharing today, only it will be less expensive and more convenient.

*Carrie lives and works in Atlanta, but doesn't own a car. She merely requests an ARS when she leaves for her mile-and-a-half trip to work, and it usually arrives within five minutes. The cost is typically $3, much less than it previously cost to take a taxi and just a bit more than the $2.50 price of MARTA, Atlanta's public transportation system. It also takes her only 5-7 minutes to get to work, compared to the 15-25 minutes it takes by public transportation, and she doesn't need to walk five minutes each way to the bus stop. After work, she* *frequently requests an AV to take her to meet friends for dinner. Overall, Carrie estimates that she spends about $125 per month for commuting and saves at least a half-hour a day compared to MARTA. For her, the benefits are obvious: it costs her $5 more per week and saves two-and-a-half hours. She estimates that she spends $200-$250 per month for all her transportation, which is only 7% of what she earns.*

## Free AV Rides

Autonomous ride services (ARS) will become so inexpensive that it will sometimes be free. For example, retail stores and restaurants will use free ARS to attract more customers.

### Complimentary Restaurant Transportation

Once ARS become available, there will be a rush to take advantage of these in changing the way services are delivered. Restaurants provide an interesting example. Many high-end restaurants offer valet parking as a service to their customers. Complimentary ARS will provide a competitive advantage for restaurants and probably won't cost more than valet parking.

*Complementary ARS will provide a competitive advantage at not much more than the cost of valet parking.*

The cost of valet parking is typically $25-$35 per hour for each valet attendant, not including tips, plus the cost to rent parking spaces in some cases. An ARS will cost $8-$15 per customer roundtrip for a typical local trip. So, the costs are reasonably similar. Initially, there will be a substantial competitive advantage for restaurants that provide a free autonomous ride service to pick up and return their customers, potentially fueling a rush to these services.

*Sam made a dinner reservation for 7 PM on Open Table at Chez Duncan, a restaurant that provides complimentary ARS transportation to and from dinner. The app knows where he is relative to the restaurant and offers free ARS. He clicks to accept, and the restaurant's ARS picks him up at 6:30. Sam usually only goes to restaurants that provide this complimentary service; otherwise, he pays for the ARS himself. He never drives to restaurants anymore so that he doesn't need to worry about parking and drinking. When he and his guests finish their dinners, the waiter notifies the ARS and the complimentary car will be waiting to take him home. Chez Duncan negotiated an excellent deal to provide ARS to its customers and found that its business increased 15% from more customers and increased wine sales.*

### Free ARS as A Promotional Incentive

Free ARS also may be an incentive for people to shop at a particular location or attend an event. For example, a shopping mall may provide free ARS to attract more shoppers. A fundraiser could give complimentary ARS to get people to participate in its event.

*The local shopping mall provides free AV transportation on Tuesdays to* *attract more shoppers. It gets 1,000 more shoppers every Tuesday and the stores in the mall each pay from $2-$20 to fund this promotion. The mall is working with its ARS provider to include paid advertising in the AVs as an additional way to pay for the service.*

## Airport and Hotel Shuttles

Millions of people traveling by air need to commute to/from and park at the airport. Parking costs are very expensive. Driving to and from the airport can be stressful. Also, many travelers stay at hotels near the airport that provide shuttle service to and from the terminals and the hotel. AV shuttle services will completely change these.

### ARS To and From the Airport

Getting to and from an airport can be very expensive. Airport parking can cost $15 to $30 per day. A car-service can be less expensive for longer trips but still costs $100-$150 each way. ARS customized for airport transportation will dramatically reduce this cost.

*Like many road warriors, Peter flies almost every week. It typically costs him $60 for airport parking, plus the cost of driving to and from the airport. A car service is $80 each way. Frequently he leaves early in the morning, arrives late at night, and he dislikes having to drive when he is tired. Now he uses an ARS designed for travel to that airport. He provides his flight information and how much buffer time he wants, and the ARS automatically picks him up in the morning. When he arrives home, he doesn't need to do anything; the ARS monitors the arrival time and is there waiting for him when he arrives.*

### Airport Shuttles

Airport hotels can use AVs to replace shuttle buses to provide much more individual service at a lower cost.

*A major hotel chain has 50 hotels at major airports. At each hotel, it previously used a shuttle bus that ran every 20-30 minutes and employed five drivers to provide the service. While customers appreciated the shuttle service, it was not very convenient since they had to wait until the next shuttle and had to stop at other terminals. They replaced the shuttles with several AVs at each hotel. Now customers have their own ride to their terminal whenever they are ready, and on top of that, it costs the hotel much less than the shuttle service. The hotel decided to purchase a small fleet of AVs identifying the hotel for advertising and promotional purposes. Even though the investment in AVs was $10 million for its 50 hotels, the return on investment was impressive, and that's not even counting an increase in revenue from guests who now stay at their hotels instead of another.*

## Transportation for Those Who Don't Drive

There are more than 50 million people in the United States who don't or can't drive. This estimate includes children, those who are disabled, and the elderly who prefer not to drive. The availability of ARS will enrich their lives and take a burden from those who now need to drive them around.

## Children

There will be some debate about the minimum age for ARS passengers. Indeed, those who are close to driving age in their early teens will be capable, but probably not young children. I envision ARS geared toward transporting children with an adequately controlled environment and continuous video monitoring so parents can watch their children.

*Jose and Brandy have three children ages 9, 12, and 14 who are all very active in sports and school activities. In a typical school week, they estimate that they have 8-12 events that their children attend. Before using ARS for their children, they felt like they ran a bus service. Now their oldest always uses the Uber ARS to get to and from sports and music practices, frequently traveling with one or two close friends. He is also very comfortable traveling by himself, and he prefers it to having mom or dad drive him because that is embarrassing socially. The two youngest children frequently go together using the Apple Children's ARS, which requires an authorized adult to be there when the kids get in and out of the car, biometric identification of the child, and continuous monitoring of the children during travel. The 12-year-old can't wait to turn 13 to be eligible to use regular ride services, and the parents can't wait for the youngest to become ten so she can travel alone in the Children's ARS. The parents estimate that they save 6-8 hours each week from not having to drive the kids. They use this time for work and shopping, enabling them to spend more quality time with the children with the time saved.*

## The Elderly

There are 16 million people older than 75 in the United States, and while many of these can drive, some of them will prefer to use ARS when it's available. For those who cannot drive any longer, ARS provides them with the freedom to come and go as they want.

*Tom and Mary are in their late 70s. Tom drove until two years ago when he had a small accident. After that, they stayed home most of the time, except for when their daughter could take them to the store or the doctor. Then their daughter convinced them to try the new ARS recently opened in their community. Now they are out and about all the time, more active than they were in many years. Mary goes shopping 4-5 times a week, before that she could only arrange to go once a week. She also takes the ARS to church activities and knitting twice a week with her friends. Tom uses the ARS for his weekly doctor's appointments, to have breakfast three times a week with friends, and for his regular Thursday night card game. On Thursday night, he typically shares an autonomous ride with two friends. Tom and Mary are so happy with the new lifestyle enabled by ARS that they convinced two friends to relocate to their community because of the availability of autonomous ride services.*

### The Disabled

ARS will provide flexible transportation for the millions of people who are disabled. In some cases, the ARS service will provide customized AVs for those who are disabled, or the person will own a customized AV.

*Bart is unable to walk and requires a wheelchair, which he is very capable of using by himself. Ironically, he was disabled in an automobile accident a decade ago. Bart has a customized AV van. Whenever he wants to go someplace, he remotely opens the garage door, pushes a button to open the back of the vehicle and the ramp extends to the ground. He then moves his wheelchair into position and pushes another button for it to lift him into the van. He then tells the autonomous van where he wants to go. Previously Bart owned a van that accommodated him, but he still needed a driver to take him anyplace.*

## Medium and Long Drives

AVs will also change the way people take long drives. In some cases, traveling by AV can be less expensive and more convenient than airline travel. Here are a few examples.

### Using an Individually-Owned AV

Some people may find that it's more cost effective to use their AV to drive long distance. Here is an example.

*Donna and Sam travel 1,500 miles every year from Boston to Miami Florida. They usually drive down in October and drive back in May, so they have their car at each location. Sometimes they would fly and pay to have their car transported. Sam looked at the economics of using an AV. He said they drive the 1,500 in three days, including two overnight stops at hotels. If they used an AV, they could do the trip in a single long day, probably 24-28 hours without stopping, except to refuel and take breaks for food. They can watch movies together on the trip and sleep along on the way. They save about $500-$600, plus two days of time, and the stress of driving. The savings offset the extra cost for an AV. Sam also compared the cost of flying home for the holidays. Flying each way cost them about $750-$900, including the cost of airport parking, plus they needed to spend another $300-$500 to rent a car. Sam figured it was about the same cost to take a day and have their AV drive them back and forth for the holidays. And they could fill the car with all their holiday gifts.*

### Family Member Medium-Distance Transportation

Here is an example of how an AV helps in a specific situation involving a family member.

*Martha's mother lives 120 miles away and doesn't like to drive. So, to get her mother to visit, she and her husband need to drive four hours round-trip to get her and then another four hours to bring her back. This trip is especially difficult on holidays since they are already busy with their kids. As a result, they don't see her mother as frequently as they would like. Now they have an AV, and they send it to pick up their mother. They usually put a snack in the car and set up one of her favorite movies for her to watch on the two-hour trip. It only took two visits for her mother to get comfortable with the AV, but now she loves it. On the first rip, Martha accompanied her mother in the AV. Now Martha sees her mother much more frequently.*

### Medium-Distance ARS from Rental Car Companies

Rental car companies will suffer from ARS replacing the need for rental cars. However, there may be opportunities for rental car companies to use AVs for medium-distance travel. Travel between two distant locations may be an opportunity for rental car companies, which is different from municipal ARS fleets.

*A formerly successful car rental company struggled to survive in the 2020's with the success of ARS. In 2023, it introduced an AV long-distance rental program. For less than $1 per mile, a customer could rent an AV for a minimum of $250. This program became an instant success, taking passengers away from the airlines. Boston to New York, or New York to DC, or LA to San Francisco cost only $250 – door to door. Even trips from New York or Boston to Chicago were cost-effective. In many cases, the travel time was less than the flight time plus the time it took to get to and from the airport. And it was more comfortable and convenient.*

*Traveling by AV can be less expensive and more convenient that airline travel.*

*Jim and Sue live in the Boston suburbs and now travel to New York a few times a year for weekends. The long-distance ARS picks them up at their home on Friday afternoon and drops them at their hotel in New York in time for a late dinner. They usually have a bottle of wine on the way to get in the mood for their weekend. On Sunday after brunch, the ARS picks them up at the hotel, and they are home by late afternoon. The round-trip costs them $500, much less than the $1,000 it would cost to fly, park and take a taxi from the airport in New York and back again. Plus, it is faster and much more convenient.*

## Business Travel

Business travelers need to suffer through regular travel to meetings and to visit clients; frequently these trips are too long to drive but inconvenient and expensive to fly. With AVs, they can now travel productively by car instead of flying. They will also find creative uses for AVs. Here are some examples.

### Travel to Business Meetings

Executives and managers frequently travel to business meetings in small groups. Some companies will purchase a few specialized AVs to transport people to and from these meetings. These AVs will lower travel costs and significantly improve productivity and convenience.

*Advanced Equity Investors is based in Boston, but its executives, and managers frequently had to travel to New York. These trips were expensive,* inconvenient, and the travel time wasn't productive. So, the company acquired three four-passenger AVs designed for business. Its executives merely schedule the vehicle for a trip to New York, and the car picks up each of the executives at home in the morning. On the trip to New York, they review their presentation and make changes as needed. Each of the business AV's has a printer so they can print client copies before arriving. The AV drops them off directly at the client office and then goes to the nearest AV parking lot to wait. When the meeting is over, they summon the vehicle to pick them up and drive them home. The company even provides some wine and cheese in the AV for the ride back.

### Salespeople

Salespeople typically travel several days a week to visit the businesses that they serve. With AVs, they can significantly increase productivity.

Brian services 400 accounts from Maine to New Jersey, and he is on the road 3-4 days a week. His company just provided him with an AV specifically designed to enable him to work while traveling from account to account. He can now do his preparation and follow-up work instead of driving, allowing him to be much more productive. He estimates that he can visit 25%-30% more accounts each week, and he not fatigued from driving so far, especially in traffic. His sales have increased significantly, and the company has gotten a fantastic return on its investment in three AVs for its salespeople.

> *With AVs, salespeople can significantly improve productivity.*

### Business Promotion

Some businesses will offer clients free AV travel to meetings as a convenience to promote their activities. Here is an example.

The Estate Planning law firm does estate and wills for wealthy clients. It acquired two luxury AVs to use for its clients. When it schedules a meeting with its clients, it offers to have them picked up and taken to the office using one of these AVs. The clients love it. They feel very special and don't need to drive or find a place to park. Its clients tell all their friends and business is up 30% since they started providing this service. It also saves the Estate Planning partners from having to travel to their clients, so that they can be more productive.

## Delivery

Package and food delivery will be revolutionized with the advent of AVs specialized for delivery. As I will discuss later, food delivery will benefit much more than package delivery. Here are what I think will be a couple of typical examples.

### Pizza Delivery

Pizza delivery will be one of the first new markets to adopt delivery by AVs. There are 70,000 pizzerias in the United States, most of who offer delivery. More than 1 billion pizzas are delivered every year in the United States. There are 16,000 Pizza Hut locations and 13,000 Dominos locations. Let's use Dominos as our fictitious example.

*Dominos was one of the first to use AV delivery vehicles for pizza delivery. Marco ordered pizza at least twice a week from Dominos. With his friends Max and Becky visiting, he ordered two pizzas from his Dominos app. They had never ordered pizza from an AV delivery vehicle, so they were intrigued by how it works. About 25 minutes later, the app notified him the delivery was 3 minutes away from his apartment. (He set the notification time for 3 minutes because that's how long it took him to go downstairs.) He confirmed the delivery, and the app sent him the "pizza passcode." They all went downstairs to see the process. The unique little "Dominos Dart" pulled up at the front door, the compartment with his pizza flashed, and he entered the pizza passcode. The door opened to the heated storage tray, he removed the two pizzas, and they went back upstairs to their apartment. "It saves me $6 for delivery between the lower delivery charge and a tip. Makes the pizza 30% cheaper!" Marco told them.*

*Pizza delivery will be one of the first new markets to adopt delivery by AVs.*

## Meal Delivery

Food delivery is not limited to pizzas. Higher-end restaurants will also offer more home delivery. Some meal delivery services are doing this already with drivers, but it becomes much more cost-effective with AV delivery, as can be seen in this example.

*Chuck and Amy regularly order meals from about a dozen local restaurants that offer AV delivery for free. Today they had another couple over for dinner. They talked about the different restaurant options and picked a restaurant, went online to the menu, and ordered on their account. Then they had cocktails, and in about 40 minutes they got a notice that their meals were waiting outside. They just went out their front door, put in their account code, and removed their meals from the warming draws. The frozen desserts were in the cooler. They prefer the convenience of dining at home more often without having to cook, and they figure that they save money because there is no delivery charge, and they don't have to pay a tip or the markup on wine with dinner.*

# Major Event Travel

Massive parking lots and incredible traffic jams are taken for granted at major events such as concerts, football games, major golf tournaments, and other sporting events. AVs may provide some solutions to these problems and make these events more attractive to attend in person.

## Big Football Games

Let's take a big football game as an example. As many as 80,000 people, sometimes more, attend a big game, creating severe logistics problems. More importantly, those attending these games get frustrated sitting in traffic, sometimes after having too much to drink. AVs, particularly ARS, can help solve these problems.

*The NFL team introduced an AV program for its games to alleviate congestion and frustration. It uses several different types of autonomous transportation. Individuals with their own AVs are dropped off in an exclusive drop-off zone, then their cars drive away to an offsite parking area. After the game, the AVs return on a rotating schedule to pick up the passengers. ARS passenger vans bring groups of 5-8 fans from their homes to the game together. They can mingle and have drinks on the way and after the game since nobody is driving. They rent the ARS passenger vans for the day and use the same van each way. And the driver doesn't need to get frustrated by the traffic. These vehicles also drop off and return on a rotating basis. Finally, conventional ARS vehicles drop off and pick up passengers from the game at a lower cost than parking fees. The vehicle drops them off and then goes to pick up a new group of passengers for the game. There is some wait time for these rides, but there is a lower price for those who agree to wait longer. This program significantly reduced congestion, eliminated the frustration of driving in traffic, and increased the interest in attendance. To accommodate the increased need for vehicles, the ARS companies dispatch the AVs in groups to the event. Some go from one major event to another around the region.*

# Chapter 2
# Introduction to Autonomous Vehicles

There is much confusion about what autonomous driving is and what it isn't. There are multiple definitions, and depending on the definition you choose, there can be different expectations and timing for autonomous driving. Historically, a lack of clarity in terminology is typical with major new emerging technologies, products, and markets. New vocabulary needs to be defined, and it takes a while for everyone to agree on what words to use. In the case of autonomous driving, many things as we currently know it will change, so we struggle for common vocabulary to describe and understand it.

I'll start by breaking autonomous driving into three simple categories: semi-autonomous, sufficiently-autonomous and fully-autonomous driving. Anyone with experience in this industry will see a deviation from some frequently used definitions. I've spent a great deal of time deciding how to categorize autonomous driving and concluded that these three categories have the highest utility. The importance of this will become clear as you progress through this book.

Currently, the most accepted definitions were created by the industry association Society for Automotive Engineering (SAE), which defines five levels for autonomous driving. I'll explain these and map them to my more straightforward definitions. I'll then go on in this chapter to describe the individual functions of autonomous driving by category, explain how an autonomous vehicle (AV) sees the world, and the discuss how the interior of an AV will be different.

## Categories of Autonomous Driving

Even though there are many degrees of autonomous driving, there are three primary categories: semi-autonomous, sufficiently-autonomous, and fully-autonomous. These categories start from nothing autonomous (what I'll call "dumb cars") to fully autonomous. Of these categories, sufficiently autonomous will be the one that will come rapidly and have the most initial impact.

## Semi-Autonomous Driving

Advanced Driver Assistance Systems (ADAS) provide the entry into semi-autonomous driving. ADAS begins to add autonomous capabilities to cars. The minimum functionality for a semi-autonomous vehicle combines three capabilities: adaptive cruise control combined with lane keeping and automatic stopping. Adaptive cruise control automatically controls the speed of the car at a level set by the driver, accelerating and slowing the vehicle based on a predetermined safe distance behind the vehicle ahead of it. Lane keeping controls the steering of the vehicle within painted lane markings on the road. While lane keeping holds the car in the lines, it doesn't turn the car around corners or sharp curves. Automatic stopping comes in variations by different manufacturers, but most of these perform well enough to stop the vehicle automatically when needed.

These three capabilities enable a car to drive comfortably on its own for long distances, primarily on interstate and state highways, but also on many secondary roads.

Currently, all semi-autonomous vehicles require the driver to be attentive and monitor the car, usually controlled by requiring hands on the steering wheel. New vehicles are also introducing cameras facing drivers to make sure they are paying attention. These restrictions are primarily to minimize the risk for manufacturers of semi-autonomous vehicles until more experience is gained. Semi-autonomous cars can go for long distances, especially on highways, without driver intervention.

I've personally driven two Mercedes with these semi-autonomous functions for thousands of highway miles, and I've also driven them semi-autonomously on regular roads, without doing much other than touching the steering wheel and making lane changes. I found that I drive very differently and more safely with these semi-autonomous capabilities. I put the car into the Mercedes Distronic Plus mode, set the speed I want such as 70 MPH, and then let the car do the driving, although still staying aware of the situation. I'm comfortable maintaining that speed or an adjusted slower speed based on the speed of the car ahead, even if other cars are speeding by and cutting in lanes. The car reacts properly. Previously, I drove like many other drivers, at a speed that was more in relation to other cars (usually a little too fast), and frequently moving from one lane to another.

Semi-autonomous vehicles are much safer. They maintain a regular speed and aren't tempted to go faster based on other traffic. They keep a safe distance behind the car ahead, and they can stop more quickly than a human driver when traffic ahead abruptly stops. Semi-autonomous vehicles apply the 80/20 rule. They accomplish 80% of the driving with 20% of the capabilities.

In the semi-autonomous category, new capabilities will be introduced progressively to increase the vehicle's degree of autonomy. These features include the ability to automatically park the car, retrieve the car from a garage, avoid pedestrians or obstacles in the road, drive autonomously in traffic jams, turn the car onto an exit ramp, etc. In other words, a vehicle becomes progressively autonomous as more features are added. Some of these features require new sensor hardware, but software upgrades may enable most of them. At some point, enough features will be added to make the vehicle sufficiently-autonomous and eventually fully-autonomous.

## Sufficiently-Autonomous Driving

Many people get hung up thinking that autonomous vehicles won't be able to do everything and go everyplace (that is fully autonomous) for a long time, concluding that

*Sufficiently-autonomous cars will come rapidly and have the most initial impact.*

autonomous driving is a long way off. While this is probably true, it is irrelevant. Sufficiently-autonomous driving only requires a vehicle to drive a predetermined route from Point A to Point B. In its simplest form, this route could be a limited route, such as taking passengers from a hotel to an airport, but it can include almost all locations and road systems in selected metropolitan areas. The essential pragmatic point in this definition is that autonomous driving can be successful without being able to drive on all dirt roads, maneuver through a difficult road construction site (instead the car will avoid it), drive during a blizzard, complete the trip by driving down a narrow alley, maneuver through a complicated intersection, etc.

Sufficiently-autonomous vehicles will be able to negotiate almost all turns, stop at traffic lights and stop signs, avoid obstacles, park themselves, etc. They will accomplish 95%-98% of most driving requirements but will achieve 100% of what is required for defined routes.

With this definition, sufficiently-autonomous vehicles will become feasible and economically successful much earlier than waiting for 100% fully autonomous driving. All too often now, autonomous vehicles are held to the unrealistic standard of 100% autonomy. This is like predicting that nobody would buy a boat unless it could travel in the high seas, maneuver down narrow rivers, and work well in low-tide on tidal rivers. People buy boats that work sufficiently in the conditions they intend to use them for.

This sufficiently-autonomous distinction is critical because, as I'll explain later, this will enable the first wave of autonomous ride services vehicles and start the autonomous driving revolution. All the primary streets and routes in a city can be mapped to the level necessary for autonomous driving, and the necessary street infrastructure can be implemented to enable seamless driving throughout the city. Passengers can request an autonomous ride service with a from and to location, and if that is a feasible predefined route, the vehicle will pick them up.

There are two subcategories of sufficiently-autonomous vehicles. Those designed for autonomous ride services will not have a driver, and the interior will be designed for the comfort of passengers. Passengers will order a ride through an app for a feasible route, and a driverless car will pick them up. AVs designed for private use will accommodate a driver when necessary. In most cases, a driver will still be required to be in the car, but the car will also be able to drive autonomously on approved routes, both with and without passengers.

### Fully-Autonomous Driving

Fully-autonomous driving will accomplish all, or virtually all, driving requirements. Licensed drivers will not be necessary. The autonomous vehicle (AV) will be in full control of all driving. Everyone in the car will become passengers. It will take some time for everyone to get comfortable with AVs, so it will take some time for fully-autonomous AVs to replace dumb cars in the United States entirely, but this will happen eventually.

## SAE Levels of Autonomous Driving

Issued January 2014, SAE International's J3016 provides a standard taxonomy and definitions for automated driving to facilitate collaboration within technical and policy domains. The six levels of driving automation span from no automation to full automation. A crucial distinction is between level 3, where the human driver performs part of the dynamic driving task, and level 4, where the automated driving system performs the entire dynamic driving task.

| SAE level | Name | Narrative Definition | Execution of Steering and Acceleration/ Deceleration | Monitoring of Driving Environment | Fallback Performance of Dynamic Driving Task | System Capability (Driving Modes) |
|---|---|---|---|---|---|---|
| *Human driver monitors the driving environment* | | | | | | |
| 0 | No Automation | the full-time performance by the *human driver* of all aspects of the *dynamic driving task*, even when enhanced by warning or intervention systems | Human driver | Human driver | Human driver | n/a |
| 1 | Driver Assistance | the *driving mode*-specific execution by a driver assistance system of either steering or acceleration/deceleration using information about the driving environment and with the expectation that the *human driver* perform all remaining aspects of the *dynamic driving task* | Human driver and system | Human driver | Human driver | Some driving modes |
| 2 | Partial Automation | the *driving mode*-specific execution by one or more driver assistance systems of both steering and acceleration/ deceleration using information about the driving environment and with the expectation that the *human driver* perform all remaining aspects of the *dynamic driving task* | **System** | Human driver | Human driver | Some driving modes |
| *Automated driving system ("system") monitors the driving environment* | | | | | | |
| 3 | Conditional Automation | the *driving mode*-specific performance by an *automated driving system* of all aspects of the dynamic driving task with the expectation that the *human driver* will respond appropriately to a *request to intervene* | System | **System** | Human driver | Some driving modes |
| 4 | High Automation | the *driving mode*-specific performance by an automated driving system of all aspects of the *dynamic driving task*, even if a *human driver* does not respond appropriately to a *request to intervene* | System | System | **System** | Some driving modes |
| 5 | Full Automation | the full-time performance by an *automated driving system* of all aspects of the *dynamic driving task* under all roadway and environmental conditions that can be managed by a *human driver* | System | System | System | **All driving modes** |

These levels are descriptive rather than normative and technical rather than legal. They claim not to imply an order of market introduction, but the order described is logical. A vehicle may have multiple driving automation features such that it could operate at different levels depending upon the features that are engaged. The system refers to the driver assistance system, combination of driver assistance systems, or automated driving system.

Fully-Autonomous driving occurs at Levels 4-5. There is some debate about the meaning and intent of SAE Level 4 as being sufficiently autonomous. There are also differing opinions regarding the need for a driver to intervene in an emergency such as an unanticipated highway situation, in an unexpected non-critical situation such as a disabled car blocking traffic, or in an anticipated situation such as driving on an uncharted dirt road.

In my definitions, semi-autonomous is equivalent to SAE Level 2 and is already achieved by several cars on the market today. Sufficiently-autonomous vehicles are achieved by Level 3-4 where the difference is that the use of autonomous driving is operational for predefined routes. This is a critical distinction in the usefulness of autonomous vehicles that is not addressed in the SAE framework. Fully autonomous is equivalent to Level 5.

## Basic Driver-Assistance Functions

Perhaps the best way to understand autonomous driving is to consider the driving functions that it performs. I'll review these progressively from basic driver-assistance functions that most people are already familiar with to those typically included in semi-autonomous driving to those necessary for fully-autonomous driving. I describe these here from a functional or user viewpoint, not a technical one. In Chapter 6, I'll review the enabling technologies.

The first set of basic driver-assistance functions are those that set the stage for autonomous driving but are not part of the semi-autonomous driving functions.

### Automotive Navigation Systems

Most people are already familiar with and comfortable with navigation systems used to position a vehicle on a visual map. It typically uses GPS to establish its position, which is then correlated to a location on a map. The Global Positioning System (GPS) was initially developed and maintained by the U.S. Department

of Defense. It uses the transmission of microwave signals from a network of 24 satellites orbiting 12,000 miles above Earth to pinpoint a vehicle's location, as well as its speed and direction of travel. Initially restricted to military use, President Reagan authorized the civilian use of GPS in 1983 after 269 passengers and crew died on a Korean airliner that was shot down when it strayed off course into Russian airspace. GPS quickly became a widely-used navigation aid throughout the world. The early versions of GPS navigation systems cost between $35,000-$70,000. Compared to their early predecessors, today's GPS devices are quite compact, inexpensive and extremely accurate.

By combining the use of signals from the satellites with interactive computerized maps in the vehicle, GPS car navigation systems can plot routes of travel to a given destination. Some GPS car navigation systems are interconnected with sources of traffic information, enabling them to automatically account for construction and congestion when determining the best route. If a driver misses a turn,

GPS car navigation systems can quickly correct for the error with an updated routing. They also can help drivers find the nearest gas station or their favorite restaurant.

Currently, standard GPS is only accurate to within approximately 5 meters or so of an object's actual location. This accuracy is sufficient to position a vehicle on a particular road in a digital map. (You may have noticed your vehicle position sometimes shifting from one road to another when turning.) But it is not accurate enough for determining what lane a vehicle is in or when to start turning. Autonomous vehicles need not only road accuracy, which lets the car know which road it's on but also lane accuracy with precision down to the meter or several feet. This level of accuracy enables the car to know which lane it's in and is accurate enough for the car to know when it has enough leeway to switch lanes or make turns securely.

GPS signals also get lost or diminished in tunnels or so-called urban canyons where tall buildings block the signal.

### Back-Up Cameras

A backup camera is a type of video camera that is produced specifically for aiding in backing up and alleviating the rear blind spot.

The design of a backup camera is unique because it flips the image horizontally so that the output is a mirror image. This mirror image enables the camera and the driver to face opposite directions. Otherwise, the camera's right would be on the driver's left and vice versa. This reduces the camera's ability to see faraway objects, but it allows the camera to see a continuous horizontal path from one rear corner to the other.

Almost 75% of the cars in 2015 model-year new vehicles include backup cameras, and all new vehicles must include a backup camera by May 2018. According to NHTSA, standard backup cameras will go a long way toward preventing injury and death, especially among children. Almost 200 people are killed each year, and another 14,000 are injured in backover accidents when drivers reverse over another person without noticing them. Most the victims are children because they are smaller and harder to see from the driver's seat.

### Telematics

Telematics is an onboard system that receives wireless information and does something useful with it. Telematics doesn't have to include two-way communication, but most do. Usually, there's an embedded cellular modem, as with GM's OnStar. Some of the telematics work can be handled by your connected smartphone, as happens with Ford Sync.

The best way to explain telematics is to describe OnStar, the original passenger car telematics systems, first announced by General Motors in 1995. It uses a cellular data modem, GPS, and a backup battery, with connections to sensors. The box goes in the back of the car, shielded from most crashes. It connects to a roof-mounted antenna that has more range than a mobile phone. The best-known feature is automatic crash notification. When a vehicle sensor reports a significant accident, OnStar sends that information to an OnStar call center, which then makes a voice call reporting the accident and location to a public-safety answering point, virtually a 911 service. At the same time, OnStar opens a voice link to the car to get more information from the occupants to inform them that help is on the way. OnStar is also used for navigation, sending a destination to the vehicle from a smartphone or web browser, or having it looked up and sent to the car by the call center. Remote door unlock is also a common function. Over time, OnStar and other services added low-overhead, high-perceived-value features such as monthly vehicle diagnostics reports. OnStar also includes data services such as weather, sports scores, stocks, movie times, and traffic information.

### Blind-Spot Detection

Traditional mirrors can help remove blind spots behind a driver, but they typically leave large dead areas on both sides of a vehicle. Blind-spot detection systems use a variety of sensors and cameras to provide drivers with information about objects that are outside their range of vision. Cameras can provide views from either side of a vehicle that allow drivers to verify that their blind-spot is clear.

Other systems use sensors to detect the presence of objects like cars and people. Some blind-spot detection systems can tell the difference between large objects like a car and smaller objects like a person, and they will alert the driver that there is a car or pedestrian located in one of the blind spots. Some systems will display a simple warning in the corner of the rear-view mirror if there is a vehicle in the blind spot.

### Surround-View Cameras

Although this is relatively new technology, I classify surround-view cameras as a driver-assistance function since it explicitly helps a driver instead of providing a function for semi-autonomous driving. A surround-view camera system offers a birds-eye view of the car from overhead and shows the car as a moving image on the car's LCD display, along with parking lot lane markings, curbs, adjacent cars, garage walls, etc.

For example, when a car backs down a driveway it shows where the car is centered. When backing out of a garage, it shows the car moving inside the garage to avoid bumping into objects. When parking in a parallel or perpendicular parking space, the driver can correctly center the car in the middle of the spot by using the surround-camera view. On most cars, the parking view comes on automatically when the car is in reverse. Or it can be activated manually to show the view when moving forward. The camera typically only works at low speeds, generally below 5 to 7 mph.

Surround-view uses multiple cameras, typically four, with software blending these together into a single video. Typically, one camera is in the middle of the front grille. Two more ultra-wide-angle cameras look down from the side view mirrors along the side of the car. A fourth is placed just above the license plate. Software blends (stitches) the four images together and inserts an image of the vehicle in the middle, making it look like there is a camera hovering above the car.

### Adaptive Headlights

Each adaptive headlight system works a little differently since they don't all perform the same functions. Some use sensor inputs to determine when the vehicle is turning. The headlights rotate with the turn, which illuminates the road in front of the car, instead of illuminating the side of the road when cornering, or shining off the road entirely, which can lead to unsafe conditions.

Others use sensors to determine when the brightness should be adjusted. This automatic adjustment saves the driver from having to operate the high beams manually, which allows for a maximum sight distance. Additionally, some can determine how far away other vehicles are and adjust the brightness of the headlamps so that light reaches them without creating glare.

## Semi-Autonomous Driving Functions

These functions enable semi-autonomous driving, sometimes called advanced driver-assist systems (ADAS). Not all functions summarized below are necessary for semi-autonomous driving, but they are increasingly included in semi-autonomous cars. In my definition, the minimum functions for semi-autonomous driving are adaptive cruise control, lane-keeping or lane centering systems, and some form of automatic braking system.

### Adaptive Cruise Control

Adaptive cruise control (ACC), sometimes called autonomous cruise control, was the beginning of semi-autonomous driving. It automatically adjusts the vehicle speed to maintain a safe distance from vehicles ahead. Initially introduced

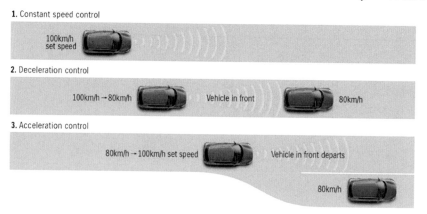

**1.** Constant speed control

100km/h set speed

**2.** Deceleration control

100km/h → 80km/h     Vehicle in front     80km/h

**3.** Acceleration control

80km/h → 100km/h set speed     Vehicle in front departs

80km/h

as cruise control, which enabled drivers to take their foot off the accelerator while the car kept a set speed, but this didn't adjust that speed when approaching too close to a car in front.

With adaptive cruise control, the driver sets a speed then the vehicle automatically adjusts its speed based on sensor information using a radar or laser sensor or a dual video camera setup allowing the vehicle to brake when it detects the car is approaching another vehicle ahead, then it accelerates when the distance separation increases.

Adaptive cruise control has already started to improve driver safety by maintaining optimal separation between vehicles and reducing driver errors. It also increases driving convenience. It is an essential component for autonomous vehicles.

## Lane-Departure Warning

Lane-departure warning (LDW) systems monitor the lane markings on the roadway and notify the driver with an alarm or vibrate the steering wheel whenever a vehicle starts to deviate from its lane. The driver can then take corrective action by steering the car back to the middle of the lane. It only works on roads with sufficient lane markings, and may not work as well if the lane markings are faded. Lane-departure warning systems are designed not to send an alert when the turn signal is on, or the brakes have been applied.

According to the National Highway Transportation Administration, 70% of all single-vehicle highway fatalities in the United States occur in run-off-road accidents. Since run-off-the-road accidents happen when a vehicle leaves its lane and drives off the roadway, lane-departure warning systems can prevent many fatal accidents. AAA says lane-departure warning systems could eliminate half of all head-on collisions.

Lane-departure warning systems have advanced to lane-keeping and lane-centering functions, but lane-warning continues to be important. These systems use several different technologies: (1) video sensors mounted behind the windshield, (typically integrated beside the rear-view mirror), (2) laser sensors (installed on the front of the vehicle), or (3) infrared sensors (mounted either behind the windshield or under the vehicle). These technologies are explained more in Chapter 6 on enabling technologies.

## Lane-Keeping Systems

Lane-keeping systems (LKS) also monitor lane markings, but they can take corrective action. A lane-keeping system can typically act to keep the vehicle from drifting. The methods that these systems use to provide corrective actions differ from one system to another. Some of the first lane-keeping systems made use of electronic stability control systems to keep a vehicle in its lane. This was accomplished by applying slight braking pressure to the appropriate wheels. Modern systems can tap into power steering controls to provide a gentle steering correction.

## Automatic Lane-Centering

Automatic lane-centering (ALC) always tries to keep the car centered in the current lane. It provides continuous control across the lane, while lane-keeping systems provide control only near lane boundaries. Controlled steering is primarily implemented through shared braking and steering control services with longitudinal control systems.

Unlike an LKS, which gently steers the vehicle back into its lane by braking the inside wheels while vibrating the steering wheel as a warning, ALC uses adaptive cruise control in conjunction with cameras and steering control electronics to keep the car centered in its lane.

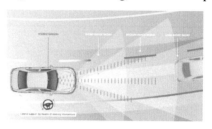

While it's not entirely autonomous, this function makes significant progress in that direction, especially considering its cameras are robust enough to perceive lanes which are nearly 1,640 feet, or 1/3 mile, away.

ALC is designed to work when the car senses drivers have their hands at least lightly on the steering wheel. In the Mercedes version of ALC, for example, the car notifies you if you have your hands off the steering wheel for more than 20 seconds. The Mercedes also uses a steering wheel icon on the dashboard to indicate if its ALC is engaged. ALC is an essential step toward vehicle automation when integrated with other vehicle control systems such as adaptive cruise control. Together these achieve Level 2 of the SAE framework.

## ALC/LKS in Snow

As discussed, ALC and LKS use a variety of sensors to read lane markers, but that's a problem when snow is covering the lane markers. That makes current ALC/LKS systems unable to function correctly. When driving in snow, human drivers who cannot see the lane markers typically make their best guess to determine road positioning based on other visible markers like curbs, signs, and other cars. Ford is teaching its autonomous cars to do something similar by creating high-fidelity, 3D maps of roads. Those maps include details like the exact position of the curbs and lane lines, trees and signs, along with local speed limits and other relevant rules. The car can use them to figure out, within a centimeter, where it is at any given moment. If the car can't see the lane lines but can see a nearby stop sign, which is on the map, its lidar scanner tells it exactly how far it is from the sign and therefore from the lane lines.

## Automatic-Braking Systems

Automatic braking systems (ABS) combine sensors and brake controls to help prevent high-speed collisions. Some automatic-braking systems can prevent collisions altogether, but most of them are designed just to reduce the speed of a vehicle before it hits something. Automatic-braking systems can save lives and reduce the amount of property damage that occurs during an accident.

Some of these systems notify the driver, while others activate the brakes with no driver input. These systems rely on sensor input, such as lasers, radar, or video data. If the sensor detects an object in the path of the vehicle, then the system determines if the speed of the vehicle is higher than the speed of the object in front of it. A significant speed differential may indicate a likely collision, in which case the system automatically activates the brakes.

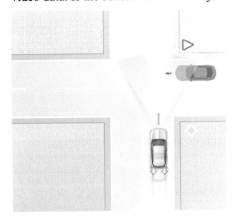

For example, Mercedes offers what it calls an Active Brake Assist with extended functionality. It warns drivers of imminent crash situations to assist them during emergency braking and, if necessary, initiates automatic autonomous braking. Active Brake Assist uses the radar sensors installed in the vehicle as well as the multi-purpose stereo camera. These enable it to detect whether the vehicle in front is slowing down, stopping or is stationary. If the system detects a risk of collision, and the driver fails to respond to a collision warning or is late responding, the system automatically initiates autonomous braking. The Cross-Traffic Function of Active Brake Assist also detects crossing traffic and pedestrians in the danger area in front of the vehicle.

## Traffic-Jam Pilot

Audi was the first to introduce specific traffic-jam functionality for autonomous driving in its A8 Traffic-Jam Pilot. When the car is in traffic, and the speed falls below 37 mph, the dash display shows a vehicle within white markings, signaling that the system is ready to take over. By pressing the Auto AI button, the car starts to drive itself. Currently, Traffic-Jam Pilot only engages the system if the vehicle is on a limited-access divided highway, has a vehicle directly in front and a line of slow-moving vehicles in adjacent lanes, and the system can identify lane markings and the edge of the roadway (with a barrier or guardrails, for instance).

In some European countries, once the system is engaged, the driver can watch TV, respond to text messages, or have a face-to-face conversation with a passenger. The entertainment and productivity features are fully integrated with the vehicle's interface so that it can adequately warn the driver if needed.

The Traffic-Jam Assist feature, available on the 2017 Audi A4 and Q7, allows for 15-second intervals of hands-off driving at slower speeds. The driver must always be alert and aware and intervene immediately when needed. Traffic-Jam Pilot will require some government approvals.

Other semi-autonomous cars may not have a specific traffic-jam function but enable adaptive cruise control with lane keeping to perform in those conditions.

### Speed-Limit Detection

Currently, some cars have speed-limit detection (SLD) to inform the driver of posted speed limits. There are two ways to do this. One uses a camera system that identifies and reads speed-limit signs using image processing. A windshield-mounted camera monitors the area in front of the car looking for road signs. A computer scans the camera image for round-shaped surfaces typical of speed-limit signs, but algorithms filter out all objects that are round-shaped but do not resemble traffic signs. The symbols are sent to the cockpit display so that the driver is always aware of the current speed limit and can adjust the car's speed accordingly.

*There are two ways to do speed-limit detection.*

Speed-limit identification in some other cars, such as those made by General Motors and Mercedes, use a roadway database in the vehicle's navigation system. The speed limit from the database is displayed in the instrument cluster. Drivers need to be careful as this is not always accurate. For example, on a recent 1,500-mile trip, I noticed a half a dozen occasions when the SLD reading was incorrect. For example, it was reduced from 55 to 45 in a 20-mile construction zone, and increased from 65 to 70 but not correct in the database.

Currently, SLD is used only to provide information to the driver and is not an autonomous function, but it will be essential to autonomous vehicles. In autonomous vehicles, speed-limit detection will determine the speed the vehicle can travel, so the speed-limit database needs to be continuously updated.

There is some discussion on whether to use SLD to set the maximum speed a vehicle can travel. It also can provide data in the car that identifies the times when the speed limit was exceeded since it has the actual speed and the speed limit. Many drivers would not find this popular, but law enforcement officials, accident investigators, and insurance companies would.

### Automatic-Parking Systems

There are different types of automatic-parking systems (APS), although they are designed to perform similar tasks. Some automatic parking systems offer hands-free parallel parking, and others only provide some helpful assistance.

Automatic parallel parking is easier for a car's computer then it is for a human. It follows a simple formula and computers are better at that:

- Pull alongside the front parked car, allowing a 3-foot gap to that car,
- Align the back tires with the front parked car's rear bumper,
- Go into reverse and start turning the wheels hard-right,
- Then start backing up slowly until reaching a 45-degree angle then stop,
- Turn the wheels hard-left,
- Then slowly continue to backup parallel with the curb,

- If needed, go backward or forward to even out the space ahead and behind.

Automatic parallel-parking only has been available for less than a decade. Automatic parallel-parking systems use a variety of sensors to determine the approximate size of the space between two parked vehicles, and then the onboard computer calculates the necessary steering angles and velocities to navigate into the parking spot safely. Early automatic parallel-parking systems had difficulty working in tight quarters.

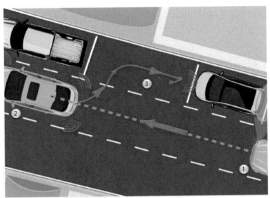

Some automatic parking systems are also capable of backing into traditional parking spaces in addition to parallel-parking. Those systems use the same technology to allow a computer to calculate the proper steering angles and velocities to park perpendicularly in between two other vehicles.

In general, most cars require the driver to change gears from forward to reverse, etc. An automatically controlled shift is coming soon on many cars. Some cars, like the Tesla, can park themselves in a garage, and back out of the garage at the owner's request.

### Over-the-Air Updates

Autonomous vehicles (AVs) are basically computers using a lot of software. Software updates are essential to AVs so that new improvements and software fixes can be rapidly and efficiently transmitted to vehicles. Over-the-air (OTA) updating is already the routine way to update cell phones and iPads, but it is not as easy for cars.

When Tesla introduced the Model S, it had an electronic architecture that enabled every line of code to change remotely. Tesla has some unique advantages in this.

The cost of transmitting a lot of software over cellular systems to millions of vehicles can be very expensive. To overcome this, Tesla uses a process for over-the-air update of its Autopilot driver-assist system that replaces only the changed code, rather than the entire file, making it much more efficient.

But there is a more significant impediment. By some state laws, auto manufacturers can't offer existing customers new features for their vehicles without the car dealerships getting their cut. This restriction contrasts with Tesla, which has done much to highlight the utility of OTA updates. Because Tesla doesn't have dealers, it isn't restricted by laws that prevent it from rolling out new features to customers without having a dealership as a mandatory middle step.

Security is also a significant issue with OTA updates to the way an AV operates, and that will continue to get a lot of attention.

Once deemed a technical challenge, remote software fixes have quickly become a priority for the world's largest car companies, even for semi-autonomous functions. Most of them are working on some solution. The lack of an industry-wide standard adds complexity for automakers, but for now, it's creating a competitive feature, with automakers vying to do the best job of satisfying regulators on security and delighting customers with new experiences.

## Sufficiently-Autonomous Driving

To become sufficiently autonomous, a vehicle needs to be able to plan and initiate turns, read traffic signs and lights, and avoid obstacles and pedestrians.

### *Path Planning*

The goal of path planning is to use the information captured in the vehicle's detailed maps to safely direct the vehicle to its destination while avoiding obstacles and following the rules of the road. Although path-planning algorithms will vary in different systems, their navigation objectives, and the sensors used, the following describes a general path-planning algorithm.

The algorithm determines a rough long-range plan for the vehicle to follow while continuously refining a short-range plan (e.g., change lanes, drive forward 10m, turn right). It starts from a set of short-range paths that the vehicle would be dynamically capable of completing given its speed, direction, and angular position, and removes all those that would either cross an obstacle or come too close to the predicted path of a moving one. For example, a vehicle traveling at 50 mph would not be able to complete safely a right turn 5 meters ahead, therefore that path would be eliminated from the feasible set. Remaining paths are evaluated based on safety, speed, and any time requirements. Once it identifies the best path, a set of throttle, brake and steering commands are passed on to the vehicle's onboard processors and actuators. Altogether, this process takes on average 50 milliseconds, although it can be longer or shorter depending on the amount of collected data, available processing power, and complexity of the path-planning algorithm.

The process of localization, mapping, obstacle detection, and path planning is repeated until the vehicle reaches its destination.

### Turning the Corner

Turing corners reliably and accurately is an essential function that AVs need to master. Currently, this can't be done with GPS-based navigation, since it's not accurate enough. More precise positioning requires detailed, sometimes called HD, mapping which enables a vehicle to know that it is 6 inches or less from a corner and can maneuver the turn.

Uber admitted that its autonomous vehicles had a "problem" with the way they handle bike lanes and that the company was working to fix a programming flaw that could see the cars making unsafe turns in the city's cycling lanes. Rather than merging into bike lanes early to make right-hand turns, as per California state law, the Uber vehicle reportedly pulled across the bike lanes, risking collisions with oncoming cyclists.

The collection of this detailed data mapping for turn management is what all the vehicles driving around with lidar towers are doing.

## Turning Left

Instructing an autonomous vehicle to turn left is one of the tougher problems. It's also a difficult problem for humans, by the way. Left-hand turns cause more than 20% of traffic accidents while right-hand turns cause only about 1%.

The AV needs to estimate the speed, distance, and timing of oncoming cars. They can do that part more efficiently than humans can do. Humans though will sometimes make judgments on when to risk cutting off an oncoming car based on intangible factors: how long the line is of cars coming toward them, how many cars are waiting behind them, and how long before the light turns red again. (Can you imagine an autonomous car being safe and polite, sitting at an intersection for 20 minutes while angry traffic builds up behind it?)

Then they need to identify how much space is required to turn and compute the proper angle for the turn. Can it complete the turn before the traffic light changes (if there is a traffic signal) and not block the intersection?

Maybe AVs will adopt the "UPS rule" which routs its 100,000 trucks in a way to avoid left-hand turns whenever possible. AVs will soon have sufficient intelligence to make these left-hand turns, and it is also possible that AV-compatible communities will implement left-hand turn traffic-signals at major intersections.

## Obstacle Avoidance

With obstacle avoidance, an AV identifies the current and predicted location of all static and moving obstacles in its vicinity. This is a continuous process of detection and identification.

The different types of obstacles are categorized in a library of pre-determined shape and motion descriptors. The AV matches obstacles it detects to this library. It then predicts the future path of moving objects based on its shape and prior trajectory. For example, if a two-wheeled object is traveling at 40 mph versus 10 mph, it is most likely a motorcycle and not a bicycle and will get categorized as such by the vehicle. The internal map incorporates the previous, current and predicted future locations of all obstacles in the vehicle's vicinity, which the vehicle then uses to plan its path to avoid the obstacle safely.

## Collision Avoidance

Automobile collision avoidance systems operate under the guiding principle that even if an impending collision is unavoidable, the right corrective measures can reduce the severity of an accident. By reducing the severity of an accident, any damage to property and injuries or loss of life are similarly reduced. Collision avoidance systems use a variety of sensors that can detect unavoidable obstructions in front of a moving vehicle. Depending on the system, it may then issue a warning to the driver or take any number of direct, corrective actions.

Most automobile collision avoidance systems draw on existing technologies. Since these systems require front-facing sensors, they often pull data from the same sensors that are used by an adaptive cruise control system. Depending on the system, those sensors may use radar, lasers, or other techniques to map the physical space in front of a vehicle. When it receives data from front-facing sensors, a collision avoidance system performs calculations to determine if there are any potential obstructions present. If the speed differential

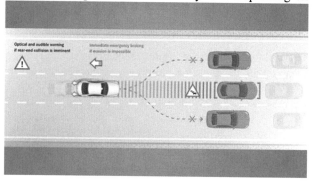

between the vehicle and an object in front of it is too high, then the system may perform a handful of different tasks. With sufficiently-autonomous vehicles, collision avoidance systems will direct corrective measures. If the system determines that a collision is imminent, it will engage the brakes rather than merely pre-charging them.

## Evasive Steering

Evasive steering helps an AV steer around an imminent crash. It uses a front-mounted camera and radar detector to monitor the surrounding traffic. If a collision is imminent, the vehicle applies the brakes. Evasive steering turns the car the exact right amount to avoid the obstacle, while also preventing the car from crashing.

Evasive steering comes into play avoiding pedestrians or other objects in the roadway. This illustration from Bosch shows a pedestrian wandering in front of a vehicle.

The cameras detect the pedestrian as well as oncoming traffic. The AV then instantly computes the evasive maneuver and initiates evasive action faster than can be done by a human.

This example, also illustrates one of the most discussed issues of AVs. What is the priority of evasive actions? Does it avoid the pedestrian with a probability of colliding with the oncoming vehicle, or does it avoid the pedestrian and crash the AV into a tree or another non-human obstacle? This example is a variation of the old trolley problem in ethics. Do you

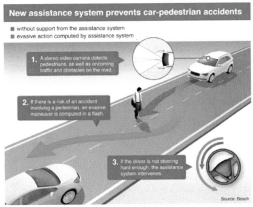

divert the trolley to kill one person or do you let it continue on and kill many? Then there are variations of this such as do you kill a healthy person to get organs to save three others? With AVs, the logic will be to try to minimize accidents in a split second, which is better than any human can do.

### On-Ramp and Off-Ramp

Once on the highway, an autonomous vehicle needs to exit the highway at the appropriate time safely. The Tesla on-ramp and off-ramp function is an example of this. Once on the freeway, the Tesla will determine which lane it needs to be in and when to exit correctly. In addition to ensuring it reaches its intended exit, Autopilot will watch for opportunities to move to a faster lane when the car is behind slower traffic. When it arrives the exit, it will depart the freeway, slow down and transition control back to the driver. In future versions, it will transition control to the next turn at the end of the exit.

### Traffic-Sign Recognition

Traffic-Sign Recognition (TSR) is a technology that enables a vehicle to recognize traffic signs, e.g. "speed limit" or "children" or "turn ahead." The technology is being developed by several automotive suppliers, including Continental and Delphi. It uses Image processing techniques to detect traffic signs based on color, shape, etc. The Vienna Convention on Road Signs and Signals signed in 1968 standardizes traffic signs across different countries. The convention broadly classifies the road signs into seven categories designated with letters A-H. This standardization has been essential for the development of a traffic sign recognition system for global use.

These first TSR systems which recognize speed limits were developed in cooperation by Mobileye and Continental AG in late 2008 on the redesigned BMW 7-Series, and the following year on the Mercedes-Benz S-Class.

Modern traffic-sign recognition systems use convolutional neural networks, mainly driven by the requirements of autonomous vehicles where the detection system needs to identify a variety of traffic signs and not just speed limits. This is where the Vienna Convention on Road Signs and Signals helps. A convolutional neural network can be trained to "learn" these predefined traffic signs and use this to determine the meaning of signs. Also, Waymo and Uber, as well as mapping companies, are generating and outsourcing traffic sign data sets for use in autonomous driving.

Traffic-sign recognition can fail to work in certain situations. In Naples Florida, a month after hurricane Irma, hundreds of street signs, including stop signs were down or bent over and would not be recognized by an autonomous vehicle. In the age of autonomous vehicles, repairing traffic signs will be a high priority, otherwise, autonomous vehicles will not be able to function safely.

### Traffic-Light Detection

Traffic lights pose a unique perception problem for AVs. A previously established map can be used to indicate when and where a vehicle should be able to see a traffic light, but vision is the only way to detect the state of that light, which may include detecting which sub-elements of the light are illuminated. This process includes distinguishing colors (red, yellow, green), as well as identifying and distinguishing colored arrows for turning. Although any vision task may be challenging due to the variety of outdoor conditions, traffic lights have been engineered to be highly visible, emissive light sources that eliminate or substantially reduce illumination-based appearance variations. A camera with fixed gain, exposure, and aperture can be calibrated to traffic light color levels.

*Traffic lights pose a unique perception problem for AVs.*

The most common failure conditions in a traffic light detection (TLD) system are either visual obstructions or false positives such as those induced by the brake lights of other vehicles. By using a map of traffic lights, the vehicle can predict when it should see traffic lights and take actions, such as braking gradually to a stop while alerting the driver, when it is unable to observe any lights.

TLD is obviously a requirement for even sufficiently-autonomous vehicles. Any road path will contain traffic lights, and the vehicle needs to reliably stop at a red light and proceed when it turns green. It's one of the autonomous driving functions that must be 100% accurate and reliable.

Eventually, traffic lights will emit an electronic signal that autonomous vehicles can interpret without seeing the light. These will also provide more information such as the number of seconds before a light change to help the vehicle pace itself better. However, reliable TLD will be needed until then.

## Emergency Vehicle Identification

Awareness and identification of approaching emergency vehicles is a specific function that must be managed by autonomous vehicles. The identification usually begins with sound detection of approaching sirens, from which the distance and then the approaching path is determined. This alert makes the car's camera sensors aware of looking for and identifying emergency vehicles, and the software guidance systems control the car appropriately such as pulling over to the side of the road.

Eventually, all emergency vehicles will have vehicle-to-vehicle communications that will notify other vehicles of their approach and request actions from those other vehicles.

## Passenger Communications

Fully autonomous cars will need a way to be able to communicate with passengers and take direction from them. For example, passengers may want to instruct the car to stop at the next rest area or market. They may want the car to change the destination or stop along the way. Passengers may also want to make inquiries of the car, such as how long will it take to reach the destination now? The car may also need to notify passengers such as a stop is coming up for refueling.

Most likely most passenger communications will be voice-based, but expect that there will also be a high-quality display in the autonomous vehicle for maps, and entertainment selection. Most likely there will be an onboard integrated entertainment system.

In autonomous ride services cars, there will also most likely be communications with a customer support center for that service. This communication network would respond to passenger problems or requests, as well as allowing the center to contact passengers.

# Special Situations

Unique situations don't occur frequently, but there are a large number and variety of them. Autonomous driving will need to address these, while sufficiently autonomous will need to either address them or avoid them by rerouting when possible. Here are some typical unique situations:

*Unique situations don't occur frequently, but there are a large number and variety of them.*

**Pedestrians not in a crosswalk**. A driver needs to determine if a pedestrian on a sidewalk intends to cross the street or just stand on the sidewalk. Likewise, if a pedestrian is jaywalking, the car needs to yield to the pedestrian. Autonomous vehicles will be very good at stopping when a pedestrian is in the way but will need to progress to determine a standing pedestrian's intentions. There are even more subtle variations of these situations. For example, it's well known in Boston that a pedestrian needs to avoid eye contact with the driver, or the driver won't stop, and the reverse that the driver needs to pretend not to see the pedestrian so the pedestrian will defer to the car.

**Traffic circles.** Traffic circles or roundabouts can be particularly challenging for drivers and even more so for autonomous vehicles. There are typically formal and informal "rules of the road" that drivers continuing past the next exit drift toward the middle and drivers trying to take the next exit drift toward the outside of the traffic circle. In Paris, the crazy roundabout at the Arc de Triomphe has something like eight circular lanes. A taxi driver demonstrated that the proper technique there is to drive perpendicular to all the circular lanes, cutting everyone off, until you get to the center, then turn perpendicular outward across eight lanes to exit. Just like Chevy Chase in European Vacation, until autonomous cars master these techniques, they could tend just to continue driving around and around the circle.

**Cobblestone roads.** Some cities such as Prague have an abundance of cobblestone roads, some with trolley tracks embedded in the road. AVs will find it very difficult to determine where they are and how to proceed without any lane markings and it's doubtful these ancient cobblestones will be painted. Old cities, such as Prague, will take a long time to accommodate AVs.

**Narrow roads.** Some older cities, especially in Europe, have very narrow two-way streets that only fit one car or bus at a time. To maneuver correctly, a vehicle must sometimes drive on the sidewalk to pass. Most likely, road changes, probably turning these into one-way streets will need to be made to accommodate AVs.

**Gated Communities.** A sufficiently-autonomous vehicle will need to enter through a gate without a passenger to activate the gate. As autonomous vehicles become more popular, gated communities will adopt practices such as automatically opening the gate for a registered autonomous vehicle using a coded electronic signal.

## How an Autonomous Vehicle Sees the World

At the most basic level, human drivers need to answer four questions: "Where am I?" (perceiving the environment), "What's around me?" (processing that information), "What will happen next?" (predicting how others in that environment will behave), and "What should I do?" (making driving decisions based on that information). Self-driving vehicles need to answer those questions, too. Waymo's Safety Report illustrates this with pictures.

### Where Am I?

Before our cars drive in any location, our team builds our own detailed three-dimensional maps that highlight information such as road profiles, curbs and sidewalks, lane markers, crosswalks, traffic lights, stop signs, and other road features. Rather than rely on GPS, Waymo's vehicles cross-reference their pre-built maps with real-time sensor data to precisely determine their location on the road.

## What's Around Me?

Our sensors and software scan constantly for objects around the vehicle — pedestrians, cyclists, vehicles, road work, obstructions — and continuously read traffic controls, from traffic light color and railroad crossing gates to temporary stop signs. Our vehicles can see up to 300 meters away (nearly three football fields) in every direction.

In this example, our vehicle has detected vehicles (depicted by green and purple boxes), pedestrians (in yellow), and cyclists (in red) at the intersection — and a construction zone up ahead.

## What Will Happen Next?

For every dynamic object on the road, our software predicts future movements based on current speed and trajectory. It understands that a vehicle will move differently than a cyclist or pedestrian. The software then uses that information to predict the many possible paths that other road users may take. Our software also considers how changing

road conditions (such as a blocked lane up ahead) may impact the behavior of others around it. The simulated imagery shown demonstrates how our software assigns predictions to each object surrounding our vehicle — other vehicles, cyclists, pedestrians, and more.

## What Should I Do?

The software considers all this information as it finds an appropriate route for the vehicle to take. Our software selects the exact trajectory, speed, lane, and steering maneuvers needed to progress along this route safely. Because our vehicles are constantly monitoring the environment, and predicting the future behavior of other

road users in 360 degrees around our vehicles, they're able to respond quickly and safely to any changes on the road. The green path indicates the trajectory through which our vehicle can proceed ahead. The series of green fences indicate that the self-driving vehicle can proceed and that the vehicle has identified the vehicles ahead and understands it must maintain certain headway.

## Interior of Autonomous Vehicles

To appreciate the differences in AVs, you need to visualize the interior, since it will be nothing like current cars. The interior of current cars is well established, and there are only minor variations. Two seats in the front and 2-3 back seats. SUVs and vans have more seating room.

In contrast, the interiors of AVs will look like a living room or den. Some will have the functionality of an office. All passengers will be able to face each other, and there may be a table in between.

In advanced semi-autonomous and individually-owned autonomous vehicles, the driver's seat will swivel, so he or she can turn around and drive the car as needed in the few times that is required. In this version of the car, the driver won't be expected to take control in an emergency, just handle those situations like unmarked roads or unusual parking.

In autonomous ride service vehicles (ARS), the only time the car needs a driver is when someone is servicing the vehicle. Limited driving capabilities will not be visible to passengers. Some ARS vehicles will have interiors for socializing; picture friends having cocktails in it on the way to dinner or a movie. Others may be customized as meeting space with a table in the middle for discussion, surrounded by chairs facing the table. Those designed for commuters may have separate quite work or entertainment spaces. In the cities, there may be very inexpensive small, pod-like AVs with a comfortable area for one or two people.

The interior for autonomous vehicles will create an entirely new opportunity for creative design. Vehicles will get customer reviews based on their interior design. There is a lot of opportunity for creativity. Most likely there will be some type of table designed in the middle of the seating for use by all passengers. There may also be drop down, pull around, or pop-up tables for individuals to use to do their work or play games. Inevitably there will be video screens built into the passenger space for groups to watch movies or do video conferencing.

Windows may be able to change from transparent to dark to make the interior

rior more private when requested. Perhaps, the windows will even be able to present a virtual reality of beautiful scenery from around the world, hiding what is outside. Most of the apparent development of AVs has focused on their autonomous functions, with little visible work done on new designs yet, although I bet that somewhere inside Apple and Waymo there are hidden design centers with prototypes of the interior of autonomous ride services vehicles.

# Chapter 3
# Benefits of Autonomous Driving

The economic and social benefits of autonomous driving are so overwhelming that it is inevitable. In fact, there may not be anything in history that provides benefits that are as significant and broad.

To start with, it will save 36,000 lives annually in the United States alone (and a half-a-million worldwide), and avoid millions of injuries. It will give the typical driver approximately 300 hours per year to do something more productive or enjoyable than driving. It will dramatically reduce the cost of transportation for everyone. And it will enable children, older people, those with disabilities to have the freedom to travel. It's difficult to find any change or technology that has had such vast benefits. And there will be other benefits as well.

## Significantly Reduce Death and Injury from Auto Accidents

Autonomous cars are safer. With an extensive array of sensors continuously monitoring everything happening around the vehicle, autonomous vehicles can anticipate and respond faster to almost any situation. They don't drive too fast, fall asleep at the wheel, run traffic lights or get distracted. Autonomous driving will reduce the approximately 11 million auto accidents in the United States dramatically.

*In the US alone, autonomous vehicles will save 36,000 lives per year, and in a decade, this is almost as many American lives as were lost in all wars in the 20th century.*

Experts estimate that approximately 90% of automobile accidents are caused by driver errors, which could be virtually eliminated by more intelligent and faster responding autonomous vehicles. Autonomous vehicles will be more disciplined and not speed or run red lights. A look at the causes of traffic accidents supports the conclusion that autonomous vehicles will reduce accidents:

**Speeding:** According to the National Highway Traffic Safety Administration (NHTSA), in 2015, 9,557 lives were lost due to speed-related accidents. Speeding was a contributing factor in 28 percent of all traffic fatalities in 2014. The NHTSA says that speed-related crashes cost Americans $40.4 billion each year.

**Drunken Driving:** In 2015, 10,265 people were killed in alcohol-impaired driving crashes. These alcohol-impaired driving fatalities accounted for 29 percent of all motor vehicle traffic fatalities in the United States.

**Running Red Lights:** More than 900 people a year die and nearly 2,000 are severely injured as a result of vehicles running red lights. About half of those deaths are pedestrians and occupants of other vehicles who are hit by red-light runners.

**Falling Asleep:** An AAA Traffic Safety Foundation study found that 37 percent of all drivers have fallen asleep behind the wheel at some point in their lives. An estimated 21 percent of fatal crashes, 13 percent of accidents resulting in severe injury and 6 percent of all crashes, involve a drowsy driver, according.

**Distracted Driving:** Distracted driving is an increasing problem. Activities that take drivers' attention off the road, including talking or texting on cell phones, eating, conversing with passengers and other distractions, constitute a significant safety threat. The National Highway Traffic Safety Administration (NHTSA) gauges distracted driving by collecting data on distraction-affected crashes, which focus on distractions that are most likely to affect crash involvement such as dialing a cell phone or texting and being distracted by another person or an outside event. In 2015, 3,477 people were killed in distraction-affected crashes, and 391,000 people were injured.

According to data released by the National Safety Council (NSC), in 2016 there were more than 40,000 traffic fatalities in the United States. A 90% reduction in auto accidents will save as many as 36,000 annually and more than 350,000 in a decade. To put this in perspective, this is almost as many as American lives lost in all wars in the 20th century. This savings in lives will be especially important for young adults. Traffic accidents are the leading cause of death for 15-29-year-olds. Almost everyone knows someone affected by tragic accidents killing young people.

In addition to the savings in lives, many people are severely injured by automobile accidents. Nearly 4.6 million people required medical treatment after crashes. A 90% reduction in auto accidents will keep almost 400,000 American people from being severely injured. Many of those critically injured take years to return to normal, and some never do.

On a global level, these numbers are even more staggering. Nearly 1.3 million people die in road crashes each year, on average 3,287 deaths a day, and an additional 20-50 million are injured or disabled. Road traffic crashes rank as the 9th leading cause of death and account for 2.2% of all deaths globally. I expect that autonomous driving on a global level will have less of an impact, as traffic accidents are more predominate in underdeveloped countries, where it will take a much longer time for autonomous vehicles to come into use.

A May 2015 study by National Highway Traffic Safety Administration estimated that the total economic cost of motor vehicle crashes in the United States including quality-of-life valuations was $836 billion. Overall, those not directly involved in crashes pay for over three-quarters of all crash costs, primarily through insurance premiums, taxes and congestion-related costs such as travel delay, excess fuel consumption, and increased environmental impacts. The real benefit, however, is that more people will continue to live and live without injury.

The estimated 90% reduction in traffic accidents will materially reduce the cost of healthcare in the United States. Traffic accidents are a significant cause of injury. More than 2.5 million Americans—nearly 7,000 people per day—entered the emergency room with injuries from motor vehicle collisions. Approximately 200,000 were hospitalized because of the incidents.

Safety improvements are already coming with semi-autonomous driving systems. A 2016 study by the Insurance Institute for Highway Safety found:

- Automatic-breaking systems reduced rear-end collisions by about 40% on average, while collision warning systems cut them by 23%.
- Blind-spot detection systems lowered the rate of all lane-change crashes by 14% and the rate of such accidents with injuries by 23%.
- Lane-keeping systems lowered rates of single-vehicle, sideswipe and head-on crashes of all severities by 11%, and accidents of those types in which there were injuries, by 21%.

Once autonomous driving becomes a significant percentage of driving, the press will begin to compare autonomous driving accidents to those caused by humans. "This accident would not have occurred if the other car was autonomous" will be reported with increasing frequency. There also will be more hyped reports of AV accidents to fuel the controversy.

One unintended consequence of the reduction in automobile accident fatalities is a reduction in the number of organs available for organ transplants.

## Reduce Driver Time

American drivers typically spend an average almost 300 hours behind the wheel annually, according to a 2016 survey from the AAA Foundation for Traffic Safety. The research finds that more than 87.5% of Americans aged 16 years and older drove cars in the previous year. During this time, a typical driver traveled nearly 10,900 miles (almost 2.5 trillion miles in total for all drivers). This driving equates to nearly 300 hours on the road per person each year, which is equivalent to seven 40-hour work weeks per year. Or stated another way, it's approximately 5% of a person's waking

*Freeing up driving time to do something better would be equivalent to seven 40-hour work weeks per year.*

hours of their life. If the typical person freed up their driving time to do something better, it would be equivalent to adding 3-4 years to their useful life.

Commuting time is a significant category of driving time that most people would like to eliminate. Experts estimate that autonomous driving will save more than 250 million hours of commuter time per year.

What will drivers do with all this free time when they no longer need to drive a car, and everyone becomes a passenger. They will be able to replace wasted driving time with increased productive working time or increase leisure or social activities.

- They could do work on their computers while commuting, participate in conference calls, or make business calls.
- They could use their time to get an online degree or take specialized courses.
- They could learn a foreign language.
- They could sleep or eat breakfast.
- They could watch a movie or read.
- They could play video games.
- Several people in a vehicle could have a meeting while being transported by the autonomous vehicle, such as preparing for the presentation they are traveling to do.
- Families could have family time together instead of the driver needing to pay attention to driving.
- Friends could use the time to talk or have cocktails (since there isn't a need for a designated driver).

It's still too early to predict how this will change working habits, society, and culture. Will companies allow employees to count this more productive work time as working hours, giving them shorter work weeks or increased vacation time? Will people use this time more for productive pursuits such as education or use it for increased leisure? Regardless, the increase in time available – 300 hours a year, 5% of waking time, the equivalent of 3-4 additional years – will be an unprecedented benefit for many people!

## Reduce Transportation Costs

Transportation costs are the second highest cost to American families. According to the US Bureau of Labor Statistics, transportation costs are approximately $9,000 annually or 14%, second to housing, which is roughly $10,000 or 16%. In contrast, food is only 10% of the total cost of living. For most families, much of the transportation costs come from owning a car. As we will see in Chapter 4, the cost of transportation will be roughly cut in half by autonomous ride services.

It's expensive to own a car that sits idle 95% of the time, and most families on average own more than one. In the United States, car ownership is almost two cars per household. Although, recent studies show that the trends per person, driver, and household, are fewer cars and fewer miles.

In 2016, according to AAA which has been studying the cost of car ownership for many years, owning and operating an average sedan costs $8,558 per year, which is equal to $713 per month or 57 cents per mile (not including fuel costs). The cost can obviously vary a lot depending on the car. Here is the breakdown:

- **Depreciation:** $3,759 per year ($313 per month). Depreciation is the rate at which a car's value decreases over time.
- **Insurance**: $1,222 per year ($102 per month). Insurance rates vary widely by driver, driving habits, issuing company, geographical area and more.
- **Maintenance**: $792 per year ($66 per month). While costs vary with the vehicle, there have been modest overall increases.
- **License/Registration/Taxes**: $687 per year ($57 per month). License, registration and tax costs vary by vehicle sales prices and state/local tax rates.
- **Finance Charges**: $683 per year ($57 per month). This is the cost attributable to car prices combined with other costs such as tax, title, license and registration fees.
- **Tires**: $150 per year ($13 per month).

In addition to calculating the driving costs for sedans, AAA determined the annual expenses for minivans and sport utility vehicles (SUVs) at $9,262 and $10,255 respectively.

Excluding the cost of fuel, the cost of automobile ownership to the average family can be $17,000 - $20,000 per year. And the average car sits idle for more than 95% of the time. Car ownership will be reduced significantly as autonomous ride services become prevalent. As is explained in the next chapter, this will be because of the advent of less expensive autonomous ride services. I estimate that the average American family eventually will reduce car ownership by 30%-40%, for an average cost saving of approximately $6,000 per year, which equates to a savings of about 10% in household spending. Collectively across all American households, this estimated savings could reach $750 billion per year.

*The cost of car ownership will be reduced by $6,000 per household and $750 billion in total.*

## Reduce Congestion and Traffic Jams

Autonomous driving can reduce traffic congestion in several ways. It will increase throughput on road systems by better synchronizing traffic. Think of how traffic congestion occurs on highways when a car slows for some reason, causing all the following vehicles to slow progressively until congestion occurs. If all vehicles are coordinated, traveling at the same speed, say 80-90 miles per hour, there would be more throughput with less congestion. Autonomously coordinated traffic passing through intersections will eventually allow the more efficient use of roads and fewer cars stopped at traffic lights with no traffic passing through in the other direction.

The reduction of traffic accidents will also reduce congestion. Accidents are a significant cause of traffic congestion, and congestion, especially sudden congestion, are likewise a cause of traffic accidents.

One of the leading causes of traffic jams is selfish driver behavior. When drivers space out and allow each other to move freely between lanes on the highway, traffic flows smoothly, regardless of the number of cars on the road. However, there is another benefit to vehicles traveling down the highway and communicating with one another at regularly spaced intervals. More cars could be on the highway simultaneously because they would need to occupy less space on the highway. The results of a Columbia University study showed highway capacity, measured in vehicles per-hour per-lane, could be increased to nearly 12,000, given a scenario in which 100 percent of the cars on the highway were self-driving and communicating with one another at 75 mph. This rate compares to about 3,000 human-operated vehicles per hour per lane. The improvement comes because the safe vehicle distance could shrink to about 16 feet for self-driving cars going 75 mph, compared to the over 115 feet necessary for safe stopping by human-operated vehicles at the same speed.

Autonomous driving will reduce unnecessary travel. Much travel is required to drive those who are unable to drive, particularly children, to and from locations. The reduction will reduce the load on road systems.

What is the estimated savings of reduced congestion? The cost of congestion is estimated to be $123 billion per year, considering: time wasted in congested traffic, fuel consumed, and the effect of traffic congestion on the environment. It also considers indirect costs, given that it is more expensive and time-consuming to transport goods or attend meetings in traffic congested cities. This estimate is based on the INRIX index that collects data from more than 180 million vehicles and devices out on the road every day using satellite navigation systems, GPS in cars and trucks, and information gathered by cellular carriers. The study found that in 2013, $78 billion resulted from time and fuel wasted in traffic (direct costs) and $45 billion was the sum of indirect costs businesses passed onto American consumers. Inrix calculated that congestion costs U.S. drivers an average of $1,400 per driver each year.

Traffic jams are particularly acute in certain cities. Inrix estimates that from now through 2026, hotspots will cost Los Angeles commuters $90.9 billion, New York commuters $63.9, and Boston $18.9 billion. The estimates mostly consider the value of drivers' time, which is based on median household income and other factors. The U.S. Department of Transportation estimates a value of $12.81 per hour for a commuter and $25.19 per hour for a business traveler. Non-business travel, like running errands, is worth $9.51 per hour. Inrix also factored in 57 cents per minute for the cost of fuel and the health and environmental cost of increased carbon emissions.

## Reduce the Cost of Accidents

The 11 million automobile accidents annually are also costly. Auto insurance is almost $200 billion in the United States. Property damage from accidents is more than $75 billion, and there are also other related costs. Legal costs from auto accidents are more than $10 billion. Emergency services cost about $1 billion, and the total estimated public spending costs on auto accidents is $18 billion, which averages about $150 per household.

As previously stated, autonomous vehicles could reduce accidents by 90%. In addition to saving lives, it would save almost $200 billion in damages and health care costs annually. It is expected to reduce insurance premiums by 40%-50%. While this is good news for drivers, it is not such good news for insurance companies.

## Reduce the Need for Parking

Many predict that autonomous vehicles will reduce the need for parking. Autonomous vehicles can return to a starting or alternative location instead of parking and waiting. They will be able to drop off passengers at the front door of their destination, park themselves somewhere less costly or return home, and come back to pick up their passengers when summoned. Drivers won't have the frustrating task of looking for a parking space because the car can do it all on its own. It doesn't matter how far away that parking space is.

Finding a parking spot can be frustrating and expensive. In the first-ever study to estimate the economic impact of parking pain, INRIX leveraged the INRIX parking database of more than 35 million spaces in 8,700 cities across 100 countries and combined this with a large-scale study of 17,986 drivers' parking behavior and experiences across 30 cities in the U.S., U.K., and Germany. It estimated that searching for parking imposes a significant economic burden on drivers. In the U.S., U.K. and Germany drivers wasted an estimated 17, 44 and 41 hours a year respectively at an estimated cost of $72.7 billion, £23.3 billion and €40.4 billion a year in these countries.

In the US, there are more than 40,000 garages and parking lots. Most of these asset owners rely on parking operators and equipment providers (that provide access and revenue control solutions) to manage their properties and maximize parking revenue. Overall parking is the United States is estimated to be a $100 billion industry with the equivalent cost to drivers. Autonomous vehicles will reduce this cost significantly.

For individual drivers, the cost of parking can be very expensive, particularly in some large cities. Parking can cost as much as $40-$50 for short-term (2 hour) parking and $400-$700 per month. For this reason, autonomous vehicles should have an earlier impact in cities with high parking rates.

In addition to the cost savings, looking for parking spots can be stressful. Almost two-thirds of America's drivers (61%) reported they felt stressed trying to find a parking spot, nearly half (42%) missed an appointment, one-in-three (34%)

abandoned a trip due to parking problems and one-quarter (23%) experienced road rage.

The reduced need for parking can also transform many urban areas. Interestingly, in the United States, there are three non-residential parking spaces for every car. The number of parking spaces required is usually set by local zoning to be sufficient for peak demand. The reduced need for parking lots and garages could transform cities. Buildings could be designed to use less land because of a lower need for adjacent parking. Parking lots could become green space and parks. City planners are already beginning to imagine the positive ways to use this additional space.

## Improve Mobility for Children, the Elderly, and the Disabled

AVs will provide new independence of travel for the millions of people who are blind, can't walk, or are older and concerned about driving. Blind people will be able to get to work just as easily as sighted people, enabling them to be more productive in the workforce. Children of aging parents won't need to take time off from work to make sure parents get to doctor's appointments, or to activities.

Further, parents won't have to worry about getting their kids to school in the morning, picking them up in the afternoon, driving them to soccer practice, or dropping them off at the movies on the weekend. Special child-safe AVs will provide that service. A parent or authorized adult can make sure the child gets into the vehicle. A video feed will enable the parent to watch their child in the car, and the car will notify the parent when the child reaches the destination.

## Save Energy

Autonomous driving could significantly reduce energy usage – or increase it, depending on your assumptions. These estimates are currently a matter of debate and uncertainty. Here are some of the arguments on each side of the issue.

The Energy Information Administration estimates that by 2050, autonomous vehicles could reduce fuel consumption by as much as 44 percent for passenger vehicles and 18 percent for trucks. However, those gains could be offset if autonomous vehicles make travel easier and cheaper for everyone and liberate shut-in populations, such as the elderly, disabled, and people too young to drive, the study says. By boosting the nation's total vehicle miles traveled, that scenario could slightly worsen fuel consumption.

The most significant potential downside of driverless cars for the environment is that AVs could increase the total number of miles traveled because it would make car travel easy and encourage car owners to make an extra trip rather than stay at home, as they might otherwise if they had to spend the time behind the wheel. Commuters might not mind living a few more miles—or even a few dozen more miles—away from work if they could do something else while the car did the driving. In already crowded cities like New York where parking is expensive, it might be cheaper for car owners to send their vehicle continually driving around the block, rather pay for a pricey urban parking space. And automated cars can safely travel far faster than human-driven vehicles — computers have much quicker reactions than even the best human drivers — which matters because fuel economy typically decreases at speeds over 50 miles per hour.

On the other hand, more appropriate sized vehicles could reduce energy consumption. People could order a much smaller and lighter vehicle for a single-passenger office commute than the current case where those same people might own an SUV because it can manage an occasional need for the extra space. In addition, once AVs avoid accidents, they might be redesigned with lighter weight materials, reducing energy needs.

Finally, most AVs are expected to use electric motors instead of internal combustion engines, although initially some could be hybrids. This should have a positive impact on the environment.

My guess is that AVs will have some positive affect overall on the environment and energy usage, mostly by accelerating the shift to electric vehicles.

# Chapter 4
# Autonomous Ride Services

The most significant change created by autonomous vehicles is autonomous ride services. Autonomous ride services will be perhaps the most significant transformation of this century. Most people eventually will use autonomous ride services instead of owning cars. There are many terms to describe this: mobility, mobility as a service, transportation as a service, personal mobility, robo-taxis (I hate that term because it envisions a robot at the steering wheel), on-demand rides, passenger economy, etc. Each of these has a particular emphasis and are not comprehensive enough. I'll use the term *ridesharing* to refer to the current services provided by companies such as Uber and Lyft, where drivers typically use their own vehicles. And I'll use the term *autonomous ride services* (ARS) to refer to the driverless evolution of this service.

*Most people will use autonomous ride services instead of owning cars.*

Ridesharing is a service that arranges one-time shared rides on short notice using a smartphone app. These trips make use of three technological advances; GPS navigation, smartphones, and social networks. Smartphone apps enable riders to merely request a ride and get paired with a driver. The two dominant players in the ridesharing game in the United States are Uber and Lyft.

Ridesharing is a large and emerging market. Based on Uber's estimated gross bookings for 2017 and its market share, the U.S. ridesharing market for 2017 was approximately $40 to $50 billion and expected to grow rapidly. SharePost projects this market will grow to as much as $600 billion per year while expanding to other segments such as short and long-term delivery. The advent of autonomous driving, substituting for individual car ownership, will solidify and increase this estimate.

There are several clear reasons why autonomous ride services (ARS) will be the first primary market for autonomous vehicles (AVs). ARS is where autonomous driving will be established first. The economics for ARS are compelling, and I will explore this with several models that explain these economics.

## The First Market for Sufficiently Autonomous Vehicles

ARS will be the first market for AVs for many apparent reasons. Even though ARS is a new concept, I expect that it will start first and snowball for some compelling reasons.

### ARS will significantly lower the cost of transportation.

People, particularly those who live in cities, are increasingly seeing the advantages of using ridesharing instead of owning their cars. A primary reason is the compelling economic savings. The average trip (although it varies widely) on Uber or Lyft costs about $13, for approximately a 4-5-mile trip (or about $2.60 per mile). For an average vehicle that's driven 15,000 miles a year, all costs of ownership added up to $8,558 a year, based on AAA's driving cost study. Including an average cost of fuel of $150 (approximately 1,000 miles per month), that's about $850 per month. So, the breakeven point economically to justify owning a car compared to ridesharing is approximately 65 trips per month or more, or roughly two per day. In cities, when you add the cost of parking, which could be $25-$60 per day in some cities, the cost of owning a car in the city can be twice as expensive, increasing the breakeven to 100 trips per month. This savings makes city commuting much cheaper by ridesharing services, which is why more people in cities are abandoning car ownership.

Autonomous ride services will be even cheaper. Later in this chapter, I model the estimated cost of ARS at $1.25 per mile. This cost would translate to $6.25 for 5-mile trip, and $8.75 for a more extended 7-mile trip. This increases the breakeven number of monthly trips to justify owning a car to 100-120, and 200 when parking fees are added. Let's look at these economics another way, by cost per mile. At the estimated ARS cost of $1.25 per mile for ARS, the cost of car ownership is better at 680 miles per month and 1,000 miles when the savings in parking fees is estimated.

The lower cost of ARS enables many people to reduce the cost of transportation by replacing both ridesharing and car ownership. ARS will be a cost saving for many people who drive fewer trips and lower miles. In particular, it will be a viable alternative to replace the second vehicle for the large percentage of families that own more than one.

### ARS will be more convenient in most cases.

The first chapter provided some insight into the convenience of ARS. Even people who own a car will use ARS frequently to go to dinner or out with friends to avoid the risks of drinking. They will use it to enable the trip to be more productive than driving. They will use it to replace taxis, ridesharing, and car services. They will use it to get to airports and to replace rental cars. The number of ridesharing trips annually in the United States is estimated at as much as 1.5 billion per year.

I expect that the use of ARS for convenience will drive much of this initial market.

## People will use ARS before buying an AV.

It will take time for most buyers to be comfortable with owning AVs. They will be nervous about the technology. It is new and will take time for people to understand how AVs work. There will be stories in the press that incite fear in autonomous driving. Many will be reluctant to invest in AVs that quickly could be obsolete like other emerging technologies, such as cell phones. When AVs become technically obsolete, they will need to be retrofitted with updated technology, and the resale value will drop. This will cause many potential buyers to wait a while. Finally, the cost of sufficiently-autonomous or fully-autonomous vehicles initially will be high, pricing most people out of the market. Initially, most autonomous vehicles will cost $80,000 - $100,000. Eventually, as technology advances, the price will come down to more affordable levels.

In contrast, most people will readily use ARS. Right from the start, ARS will be comfortable, convenient, reliable, and inexpensive. People will want to try it, and when they do, they will continue using it. ARS will be able to take the best advantage of sufficiently-autonomous vehicles, and riders will be able to get to where they want to go reliably.

## ARS can be introduced by municipality.

When you consider autonomous driving technologies, the ability to develop extremely detailed visual maps will be an early challenge. These detailed visual maps include all traffic signs, traffic lights, trees, curbs, lane configurations, trees, crosswalks, etc. This is discussed more in the chapter on autonomous driving technologies.

A significant advantage of ARS is that it can do very detailed mapping in a defined geographic area, the city or municipal area where it provides the ride service. When a customer requests a ride with a designated from and to location, the ARS system will determine if that is a feasible autonomous ride services route. ARS companies can enter a geographical area by doing the detailed mapping for that area and providing the number of autonomous vehicles required.

Additionally, autonomous driving will be embraced earlier by some municipalities. These will support autonomous driving by upgrading the road systems and infrastructure, and maintaining it accordingly. There will be some initial challenges getting autonomous vehicles to work efficiently in snow storms. The early solution to this could be that ARS will merely avoid these geographical areas in their initial rollout.

## ARS vehicles will be more comfortable.

Early versions of privately-owned AVs will need to accommodate human drivers because there will be areas where autonomous driving will not be feasible. However autonomous vehicles for ARS will be custom designed for passengers only. They will be designed differently than privately-owned autonomous vehicles.

Picture a small two-passenger pod-like vehicle with comfortable seating in a vehicle half the size of a typical car today. Or a four-passenger luxury vehicle designed like a living room with comfortable seats facing each other around a coffee table with a bar, music, and screen for movies and entertainment. There would still need to be some way to operate the vehicle for service and maintenance manually, but this could be minor and inconspicuous.

The interior design for passenger comfort will even become a point of competitive differentiation. With these custom designs, autonomous ride services vehicles will be more comfortable and more luxurious than typical vehicles and other privately owned autonomous cars, in addition to being less expensive.

## ARS is the real strategy of ridesharing companies.

The real objective of the current ridesharing companies like Uber and Lyft is to provide autonomous ride services. Ridesharing is merely the prelude to autonomous ride services, just as Netflix DVD subscriptions was the prelude to Netflix video streaming.

The strategy of ridesharing companies is to develop the market, get customer loyalty, then replace drivers with autonomous ride services. With the current ridesharing business model, they are losing a lot of money, yet they keep raising more capital at very high valuations to fund these loses. Based on estimates by SharePost, ridesharing companies have raised more than $25 billion in private capital since 2010. Today, the top-5 ridesharing companies – Uber, Didi-Chuxing, Lyft, Ola, and Grab – have a combined market capitalization of roughly $120 billion (based on the most recent round of investment valuations).

It's unlikely that the contractor-drivers using their own car model of ridesharing will ever be profitable. Besides, there is a lot of aggravation associated with it such as driver screening and liability. But these companies are developing the market by converting riders to ridesharing from taxis and car ownership. They are overcoming local licensing and tax obstacles and objections of taxi drivers. They are building brand names. They are creating loyal customers.

Investors know that ridesharing is not a profitable model until autonomous vehicles replace drivers. And then it will be an enormous and highly profitable business. They know this is the end objective and because of this there will be a lot of pressure for the current ridesharing companies to accelerate autonomous ride service as soon as possible.

## Technology companies will enter the ARS market.

Apple and Waymo (Google) have their eyes set on this significant opportunity for AVs, but they have critical distribution obstacles to overcome to enter the retail market selling cars to consumers. They don't have the sales channels to sell cars to customers. Tesla as a pioneer has had and still has obstacles to overcome selling cars directly to consumers. In most states in the United States, it is prohibited from selling cars directly to customers by archaic laws requiring that a manufacturer must sell only through an independent dealer.

These technology companies don't have the manufacturing capabilities to build autonomous vehicles for consumers. Although they could subcontract the manufacturing, they do not have service capabilities to provide to consumers. Consumers require adequate service, and in fact, may expect exceptional service. Putting this infrastructure in place would require time and money, with deployment time being the most significant obstacle. Finally, they would need to overcome buyer skepticism that they can be successful long-term in the market. There could be long delays for these companies to get to the necessary volume levels.

So, they will not enter the retail market for AVs; instead, they will enter the ARS market. Most of these are not obstacles in the autonomous ride services market. Customers using autonomous ride services won't care about buying or servicing cars, or if that car will last long term. These technology company entrants can provide the autonomous driving fleet in selected areas. Servicing and maintaining a fleet serving a municipal area is a solvable problem (think subcontracting to Hertz or Avis). Manufacturing hundreds of thousands of vehicles can be outsourced to contract manufacturers or the traditional car manufacturers.

These two companies have invested billions in autonomous driving, have the capital needed to jump into the ARS market, and are motivated to take advantage of this lucrative market. Outside of the United States, there may be other technology companies with the same strategy.

## Traditional car manufacturers are investing in ridesharing.

The traditional automakers are also investing in ridesharing with an eye toward using this as a platform for AVs. GM invested a half a billion in Lyft in 2016, while planning to work with Lyft on autonomous driving and also offer its own ARS service. GM also announced it was launching a personal mobility brand named Maven and also GM Cruise. GM believes it is uniquely positioned to provide the high level of personalized mobility services its customers. Maven started by offering its car-sharing program to more than 100,000 people in Ann Arbor, Mich. with GM vehicles located in 21 parking spots throughout the city. GM intends to begin offering its own ARS by 2019.

Toyota invested an undisclosed amount in Uber. Toyota is interested in exploring the future of transportation with Uber. The companies have entered into a memorandum of understanding (MOU) to explore collaboration, starting with trials, in the world of ridesharing in countries where ridesharing is expanding, taking various factors into account such as regulations, business conditions, and customer needs. Also, Volkswagen invested $300 million in the New York-based ridesharing startup Gett. Ford has stated that its 2021 autonomous vehicle will be exclusively for ARS.

## ARS will create new market segments.

There will be new services created by ARS. Think about this for a minute. I discussed some of this in an earlier chapter on new transportation alternatives for the disabled, elderly and children.

Shuttle buses with drivers will be replaced by autonomous shuttles early in the new market. Shuttles such as those from Navya can use direct and straightforward preprogrammed routes, so they can be brought to market earlier than autonomous vehicles that need to have the sophisticated mapping for much broader regions. There are a lot of shuttle buses used for hotel shuttles from airports, car rental shuttles, college campus shuttles, etc. In this market segment, the autonomous shuttles will be bought by a hotel, airport, college, etc. to reduce cost by eliminating drivers.

Paratransit is another market segment for autonomous ride services. An example of this is The Ride, the Massachusetts Bay Transportation Authority's paratransit program for people with physical, mental or cognitive disabilities that make it difficult or impossible to ride the MBTA's fixed-route bus, train, and ferry services. These shuttle services cost almost $100 million a year for Boston alone. Individual trips cost estimates are $45-$50 each. An autonomous ride service could bring this down to $8-$12 per trip on average, saving Boston alone more than $70 million per year.

The new markets are numerous.

### Capturing metropolitan markets will be like a "land rush."

I believe that companies will aggressively launch ARS in new markets to beat others to capture these markets. ARS requires a significant physical investment to create a presence in a local market. $50-$100 million will be a typical investment to establish a presence in each metropolitan market.

Once a company establishes a presence in a local metropolitan market, it will discourage other companies from investing in the same market because the chance of success will be lower. The other company will just move into another market. First-to-market has a new meaning here. The company that follows will have less potential because it will need to share the market.

This incentive will create a "land rush" effect. The primary competitors will rush to new metropolitan markets to capture that market before others. The result of this will be to accelerate the growth of ARS even faster than it usually would be.

## The Economics and Business Model of ARS

Although current ridesharing services and the future autonomous ride services (ARS) perform primarily the same task, their economics and business models couldn't be more different. Ridesharing today is a labor-intensive contractor model with no investment in cars or even employees. Drivers are all contractors using their own cars. The only key to success is the app people use to get a car, along with a sufficient critical mass of contractors to provide enough coverage for people to use the app.

The autonomous ride service of the future is a very different business model. To start with, it will be extremely capital intensive. The ARS company will need to purchase its autonomous vehicle fleet or partner with someone else who invests in the fleet and manage its utilization efficiently. Unlike ride sharing, ARS will be a very profitable business.

Let's look at the approximate economics of this business: pricing, utilization, operating costs, and return on investment. The numbers are so big, that "back of the envelop" estimates can give us a reasonable approximation of the potential.

## Pricing

Autonomous Ride Services (ARS) will offer a variety of pricing plans and programs, enabling them to fit different market needs, manage capacity utilization and achieve competitive advantage. Here are some potential pricing plans:

- **Basic Rates** – ARS rates will be set by mile traveled, by time, or by a combination of the two. Long-distance rides may offer lower prices.
- **Off-Peak and Surge Pricing** – Rates may vary to improve the utilization of the autonomous vehicles (AVs) in an ARS fleet. When demand is higher, prices will be higher. When demand is lower, rates will be lowered to encourage utilization during those time periods.
- **Promotional Rates** – It's very likely in some markets that there will be intense price competition with many special promotional programs. I also expect that there will be aggressive promotional programs to get people to use ARS for the first few rides.
- **Wait Time Charges** – Some ARS companies may offer a lower wait time charge, particularly during off-peak hours. This will enable customers to pay a few dollars per hour to keep the same AV waiting for them while they run errands, have a doctor visit, etc.
- **Subscription Programs** – I expect several versions of subscriptions programs to be available. These will be monthly or annual programs that will enable customers to get lower rates and preferred availability of services.
- **Pre-Scheduled Rides** – Unlike current ridesharing services, ARS will be able to be pre-scheduled for pick up. This is particularly useful for people who need a ride to the airport early in the morning. Pre-scheduled trips may have a scheduling fee.
- **Cleaning Fees** - One of the challenges for ARS will be cleaning in between rides. Most will have an additional cleaning fee associated with special cleaning after a trip that goes beyond typical usage.

Let's make some assumptions on pricing. An average Uber trip in 2015 was 6.4 miles, up from 5.7, with the expectation it will continue to increase as ridership increases. The average trip took 14 minutes. The average fare for an UberX trip is approximately $14.00 - $15.00, or roughly $2.25 per mile or $1 per minute. Uber Select is about twice as expensive, almost $30 per trip, for a nicer vehicle. A Guggenheim analysis estimated the average ridesharing trip cost Uber and Lyft $2.75 per mile. Others have an estimate of $2.50 per mile.

Guggenheim estimates for ARS is $1.50 per mile by 2020. In the business model to estimate revenue per ARS vehicle, I'll use $1.25 per mile as an overall average, but also show the estimated revenue for $1.00, $2.00, and even $0.50 per mile. The first step in the complete forecast for ARS is to estimate the potential revenue per ARS vehicle. The next is to estimate the gross profit per vehicle, including the typical costs for each vehicle. I expect that ARS vehicles will be managed as autonomous fleets by metropolitan area. So, the final step is to model the profitability of different sized fleet operations.

## Revenue Per ARS Vehicle

Revenue for each ARS vehicle could be very impressive. Using the following model, I project it at approximately $150,00 per-year per-vehicle using a conservative estimate of $1.25 per mile, almost half of the current price of Uber ridesharing.

**ARS Estimated Revenue Per Vehicle**

|  | Base Case | Minimum Case | Maximum Case | Longer Trips | Price Competition |
|---|---|---|---|---|---|
| Trips per Day | 50 | 30 | 75 | 20 | 70 |
| Miles per trip | 7 | 7 | 7 | 20 | 7 |
| Miles per Day | 350 | 210 | 525 | 400 | 490 |
| Trips per Hour (20 hrs/Day) | 2.5 | 1.5 | 3.75 | 1 | 3.5 |
| Average speed per hour | 30 | 30 | 30 | 60 | 30 |
| Time Used per Day (minutes) | 700 | 420 | 1050 | 400 | 980 |
| Utilization per day | 49% | 29% | 73% | 28% | 68% |
| Price per Mile | $1.25 | $1.00 | $2.00 | $1.00 | $0.50 |
| Price per Trip | $8.75 | $7.00 | $14.00 | $20.00 | $3.50 |
| Revenue per Day | $438 | $210 | $1,050 | $400 | $245 |
| Days Used per Year | 350 | 350 | 350 | 350 | 350 |
| Miles per Year | 122,500 | 73,500 | 183,750 | 140,000 | 171,500 |
| **Estimated Annual Revenue** | **$153,125** | **$73,500** | **$367,500** | **$140,000** | **$85,750** |

The model assumes a 7-mile average trip and approximately 2.5 trips per hour, similar to the current number of trips per hour today for Uber drivers. Assuming 20 hours of operating time per day, this translates into 350 billable miles per day and a little less than 50% utilization. The estimated price per trip is $8.75, which is much less than the current $14-$15 for a comparable Uber trip.

The model also shows a minimum and maximum case. The minimum case assumes $1 per mile pricing and less than 30% utilization but still results in almost $75,000 in revenue per vehicle. The maximum case assumes $2 per mile pricing and nearly 75% utilization and results in more than $360,000 in annual revenue per vehicle. There is also a price competition model, which is very possible if the market gets competitive. It assumes $0.50 per mile for an average of $3.50 per trip, with higher utilization. This still results in $85,000 per vehicle, which shows the potential of very low-cost ARS with significant revenue possible.

## Utilization

Unlike ridesharing, the usage of an ARS vehicle is not limited by driver availability so that it can drive many more miles and many more trips. Uber drivers average about 2.5 trips per hour, although that varies widely by time. Let's assume an ARS vehicle does 2.5 trips an hour for 20 hours a day, allowing for refueling time and slow business between 1 AM and 5 AM. This is only a little over 50% utilization in an hour and less than 50% per day. Over time, artificial-intelligence driven scheduling could optimize autonomous ride services vehicles with utilization of greater than 80%.

At the 50% utilization, this would translate to more than 120,000 passenger miles per-vehicle per-year, plus miles required to position a car to pick up a passenger and return to a central location after drop off. The performance of individual vehicles will vary by different companies and different areas, and that variance will become competitively significant. These metrics will become better understood as the ARS market become a reality.

## Vehicle Operating Costs and Gross Profit

The operating costs for an ARS vehicle include fuel (gas or electricity), maintenance, insurance, and taxes.

### ARS Estimated Gross Profit Per Vehicle

| | Base Case | Minimum Case | Maximum Case | Longer Trips | Price Competition |
|---|---|---|---|---|---|
| **Estimated Revenue per vehicle** | $153,125 | $73,500 | $367,500 | $140,000 | $85,750 |
| | | | | | |
| Passenger Miles per Year | 122,500 | 73,500 | 183,750 | 140,000 | 171,500 |
| Non-Passenger Miles per Year | 36,750 | 22,050 | 55,125 | 56,000 | 51,450 |
| Total Miles per Year | 159,250 | 95,550 | 238,875 | 196,000 | 222,950 |
| | | | | | |
| Operating Costs | | | | | |
| Fuel | $6,370 | $3,822 | $9,555 | $7,840 | $8,918 |
| Insurance | $3,000 | $3,000 | $3,000 | $3,000 | $3,000 |
| Cleaning & Mainenance | $20,000 | $12,000 | $30,000 | $15,000 | $20,000 |
| Taxes | $3,000 | $3,000 | $3,000 | $3,000 | $3,000 |
| Depreciation | $20,000 | $15,000 | $25,000 | $20,000 | $20,000 |
| Total Operating Costs | $52,370 | $36,822 | $70,555 | $48,840 | $54,918 |
| | | | | | |
| **Gross Profit per Vehicle** | $100,755 | $36,678 | $296,945 | $91,160 | $30,832 |
| | | | | | |
| Investment per vehicle | $100,000 | $100,000 | $100,000 | $100,000 | $100,000 |
| ROI per Vehicle | 101% | 37% | 297% | 91% | 31% |

Assuming electric cars, which is most likely, the fuel cost is currently about $0.033 per mile. The typical autonomous vehicle with the experience above would incur approximately 160,000 miles per year, including 30% additional miles to reposition and send a car. Assuming a little higher cost of $0.04 per mile, fuel costs for that vehicle would be a little more than $6,000 per year.

Cleaning and maintenance costs will be relatively expensive since the car is being driven 160,000 miles annually. Let's assume an average cleaning and maintenance cost of $20,000 per year. Insurance costs should be reasonable since the autonomous vehicles should have very few accidents, but let's conservatively estimate $6,000 per vehicle, although this will go down in the future. Also, there will most likely be city, state, and federal taxes on this new industry – let's assume $3,000 per-vehicle per-year. This is an estimated operating cost of approximately $32,000 per ARS vehicle per year before depreciation of $20,000.

Finally, considering the miles driven, the life of an autonomous ride service vehicle will be relatively short, let's say four years and add $20,000 per year in depreciation, assuming a $100,000 initial cost for each vehicle and some residual value. As you can see, the ROI per vehicle is very high, even in the minimum and price competitive cases.

### Typical Municipal ARS Fleet Operation

ARS will be provided by fleets of AVs that serve a designated metropolitan area. Each of these ARS fleet operations centers will offer a range of services: oversee the dispatch of AVs in the fleet, intervene directly in customer problems and issues, and track the location of all vehicles 24/7. Also, they will schedule or provide regular maintenance, manage the cleaning of all vehicles, upgrade AVs with new technology, manage the size of the fleet, work closely with local authorities to ensure the services work smoothly and follow regulations, and maintain the detailed mapping and geo-fencing of the municipal area. Some municipal fleet operations will also be profit centers managing marketing and promotion, setting prices relative to competition, and managing personnel.

A typical fleet operations center staff will be relatively small, probably eight people for a fleet of 200 cars, varying with smaller and larger fleets. There will also be rent expense for fleet parking. Vehicle cleaning costs were estimated in the per vehicle costs. In addition, there will be information technology, administration, and marketing expenses. A 200-car fleet would have an estimated revenue of approximately $30 million per year and an incredible expected profit of $20 million annually. A 200-car fleet might require an investment of $20-$25 million, but the return on this investment would be 70% annually.

This exceptional return on an ARS fleet operation has two implications. First, prices may go much lower with competition, but this would also grow the market faster. That's why the price competition model bears some consideration. At $.50 per mile and $3.50 per trip, a fleet operations center would still generate $17 million in revenue and profit of more than $5 million, which would still be a decent return on investment of 20% annually.

The second implication is that the potential profitability will create the "land rush" described previously. ARS companies will try to rush in and "own" as many municipal markets as they can.

Fleet operations center profits would be aggregated by the company that owns the center, and this would be used to fund corporate marketing, R&D, and administration.

How many ARS AVs will be required in a metropolitan market. Let's look at a small metro market with a tourist component. Collier County (which is primarily Naples) Florida has a permanent population of about 350,000 people with an additional 1.8 million visitors during vacation season. Collier County is about 2,000 square miles or about 45 miles by 45 miles, so an average trip of 7 miles or so is reasonable.

### Economics of ARS Fleet Operations Center

|  | 100 Vehicles | 200 Vehicles | 500 Vehicles | 1,000 Vehicles |
|---|---|---|---|---|
| Estimated Revenue per vehicle | $153,125 | $153,125 | $153,125 | $153,125 |
| Gross Profit per Vehicle | $100,755 | $100,755 | $100,755 | $100,755 |
| Revenue per Center | $15,312,500 | $30,625,000 | $76,562,500 | $153,125,000 |
| Gross Margin per Center | $10,075,500 | $20,151,000 | $50,377,500 | $100,755,000 |
| | | | | |
| **Fleet Operations Costs** | | | | |
| Operations Staff | $240,000 | $400,000 | $560,000 | $800,000 |
| Fleet Parking./Facilities Rent | $50,000 | $100,000 | $200,000 | $300,000 |
| Information Tech. | $50,000 | $70,000 | $80,000 | $100,000 |
| Administration/Other | $50,000 | $60,000 | $70,000 | $80,000 |
| Marketing& Promotion | $25,000 | $50,000 | $100,000 | $250,000 |
| Total Operations Expense | $415,000 | $680,000 | $1,010,000 | $1,530,000 |
| | | | | |
| **Fleet Oper. Center Profit** | **$9,660,500** | **$19,471,000** | **$49,367,500** | **$99,225,000** |
| **Profit Margin** | 63% | 64% | 64% | 65% |

Let's further assume that in the first few years, 20% (70,000) of the population uses autonomous ride services and that they use it an average of four times per month per person (some rides will include multiple people) so that there would be about 280,000 autonomous ride service trips per month. This is a relatively conservative estimate that could increase significantly as ARS becomes more accepted. Using our trip estimates, this would require approximately 200 autonomous vehicles in Collier County to provide this service for residents.

Tourists would probably have a much higher use of ARS since they would use this instead of rental cars. There are 1.8 million tourists per year. If the tourists travel in groups of two, stay for five days, and half of each group use ARS once per day on average, that would equate to an additional 2,250,000 trips or about 375,000 per month over a six-month period. This would require an additional of approximately 250 autonomous vehicles during the tourist season.

The 250 seasonal ARS vehicles can be deployed as needed from other municipal fleet operations. Large ARS companies will redeploy vehicles as needed. The vehicles would merely drive themselves from other centers at the start of the busy vacation period and then return afterward.

## American Market

The potential size of the ARS market can be substantial. In chapter 10, I make projections of the ARS market size over time. I estimate the market for ARS in the United States alone could be $750 billion in 2030, and potentially $150 billion in 2025.

To put that in perspective, the smartphone market in the US is approximately $55 billion annually. Apple currently has $200 billion in global revenue, Google has $90 billion, and GM has $166 billion. This is also a reasonable estimate relative to the size of the auto industry of $1.6 trillion, plus the value of other industries such ridesharing, taxis, fuel, service stations, etc.

However, the capital investment to get into this market will be incredible. The market will initially evolve by city-by-city or region-by-region. The ride services company with the best coverage will most likely get a disproportional share of the market, just as it is with the airlines. A 100-car market will require approximately a $10 - $12 million capital investment. A 1,000-car market will require a $100 - $120 million investment. A national roll-out of 1 million vehicles would be a $100 to $120 billion investment. The entire American market could eventually require 3-4 million ARS vehicles for an estimated capital investment of $300 billion to $450 billion.

It's not yet clear who will make this investment in massive ARS fleets. Some of the ARS companies, like Uber, Google, and Apple may own their own fleets. Google and Apple have the capital to do it. Others, like Lyft, may use traditional car manufacturers to make this investment because they already have similar investments in leased vehicles.

## Implication on the auto manufacturing industry

In a later chapter, we will examine the potential strategies of the many different players in this industry. However, with these estimates fresh in our minds, it's worth noting how autonomous ride services will impact the traditional auto industry. Every new autonomous ride services vehicle will replace approximately ten personal vehicles. The reason for this multiple is that the average car sits idle most of the time and is used only 5% of the time, but the average ARS vehicle will be used approximately 50% of the time. 150,000 miles per year compared to 13,000 per average individually owned vehicle.

This transition will take some time; it will be profound and clearly can be anticipated. In 2016, approximately 72 million cars were produced worldwide. If autonomous ride services achieve a 10% penetration worldwide, with 2 million vehicles, then approximately 20 million traditional vehicles will be replaced annually.

The autonomous ride services companies will most likely use some traditional auto manufacturers as contract manufacturers. Some may merely retrofit conventional car designs in the beginning. Others will build high a volume of specially designed vehicles to their unique specifications. Can you imagine Apple using a traditional car design? Mostly likely this will be welcome, but low-profit, business used to offset the reduction of 25 million cars.

# Chapter 5
# Trucks, Delivery Vehicles, and Buses

Autonomous passenger vehicles are the primary focus of this book, however autonomous trucks, delivery vehicles, buses and specialty transportation are also going to transform the services they provide. In fact, the use of autonomous capabilities may happen early in some of these segments.

There are many segments within this category. It includes long-haul trucking, package delivery, busing, and specialized transportation. Each of these segments has unique needs addressed by autonomous vehicles, but some of these needs will benefit more than others. Autonomous trucking, especially long-haul trucking will see some significant benefits, while package delivery will be more difficult because someone needs to remove the package from the truck. In contrast, autonomous food delivery will be very successful because someone is always waiting for the food. Autonomous shuttle services will become very popular, but autonomous buses may not. We will also look at specialty vehicles, particularly mail delivery and military uses.

## Autonomous Trucking

The trucking industry has annual revenue of nearly $200 billion, but it is a fragmented industry. The 50 largest companies account for less than a third of the market. Efficient operations are critical, and autonomous vehicles can provide enormous benefits. Seventy percent of the industry is long-distance trucking, and the remainder is local.

Full-truck-load shipments of general merchandise are the largest part of the market. A trucking customer typically loads a full trailer, the trucker transports the container, and a receiver unloads the contents. Shipments are usually delivered within a couple of days, depending on the distance.

The full-truck-load long-haul market will be one of the first impacted by autonomous trucking because autonomous trucking can significantly reduce these costs. Currently, a full truckload from L.A. to New York costs approximately $4,500, with labor representing 75% of that cost. Autonomous trucking will be able to eliminate most of this labor costs, both reducing shipping costs and increasing profitability. Eventually, driver-based long-haul trucking will become non-competitive. With autonomous trucking, trucks can drive nonstop (except for fueling) and take advantage of the optimal cruising speed.

Autonomous trucking will enable much faster delivery. Currently, it is legally mandated that for every 11 hours of driving a truck driver must take an 8-hour break. With fully autonomous trucks, there won't be a requirement for the trucks to stop for these breaks, allowing them to make their deliveries faster and improve lead times for companies across the United States. Let's take a typical 2,600-mile trip that takes 55 hours of driving. With four required 8-hour breaks and time for bathroom and food stops, the total delivery time would be approximately four days (96 hours). An autonomous vehicle could make the same trip in less than 60 hours, assuming refueling stops, and that is almost 40% less time. Delivery time is critical in long-haul shipments, and 2 ½ days is a significant competitive advantage over four days. There is also another considerable advantage: truck utilization. With shorter delivery times, an autonomous truck fleet could be 40% smaller and make the same number of deliveries.

I expect that autonomous truck staging areas near major highways will facilitate the process. An autonomous truck could be loaded and driven to a staging area near a highway, then set to operate autonomously to the destination, stop at another staging area next to a highway near its destination, and then driven by a driver to the final unloading destination a few miles away.

In the early years of autonomous trucking, drivers will still be on board to drive as needed at the beginning and end of each trip and to refuel the trucks, but they will be able to sleep for long periods of time while the truck is self-driving. This method will provide a significant portion of the benefits by making deliveries faster without the need to stop the truck for driver-rest periods, significantly increase truck utilization, and reduce the cost of drivers.

Some trucking industry experts see great potential in utilizing platooning systems for cost savings and achieving more efficiency in freight hauling. One autonomous truck can lead a platoon with two or more trucks following close behind, taking advantage of the aerodynamic efficiency.

Volvo successfully demonstrated on-highway truck platooning in California. Daimler AG's truck division is following a similar path, testing platooning technology on U.S. roads. The German company's U.S. division gained approval from Oregon's transportation regulatory agency after completing a successful trial run in the state. Peloton Technology is developing a platooning system that will make it easier for trucks to travel within a platoon. That means they'll save significant volumes of gasoline and diesel typically consumed by work trucks. Peloton's system uses cameras, sensors, and networking equipment for trucks to communicate with each other.

While truck platooning has potential benefits, I don't see these as being as significant as the other cost reduction opportunities in long-haul trucking. I also anticipate pushback from passenger-car drivers who may be intimidated by long platoons of trucks speeding down the highways.

## Autonomous Package Delivery

Package delivery, sometimes referred to as the last-mile delivery, will benefit from autonomous vehicles, but the benefits won't be as high as they are in other segments. The reason the benefits are not going to be so great is that the recipient is not always there to receive a package, so someone needs to hand-deliver the package from the truck to the recipient or place it correctly by the door. Currently, there are two methods under consideration for using autonomous vehicles for package delivery.

Using a semi-autonomous ground vehicle, a delivery person is still required but could replace driving time to more efficiently take care of sorting packages or smaller administrative tasks, e.g., scanning or announcing arrival while the vehicle does the driving. These advantages need to compensate for higher investment costs, as autonomous ground vehicles are likely to be more expensive than regular trucks, at least initially.

DHL is planning to test self-driving delivery trucks that follow a delivery person along the route to deliver packages. The person wouldn't need to get back into the truck between dropping off packages; they would remove packages from

the rear of the truck. The company has not revealed how the truck will identify and accurately follow its delivery people.

There can also be efficiencies on better routing and deployment of delivery vehicles. A recently loaded truck could drive itself to the delivery person and replace a truck that is empty, for example. UPS estimates that being able to cut a single mile from a driver's daily route saves the company up to $50 million each year. On any given day, UPS has around 66,000 drivers out on the road.

Autonomous vehicles with lockers could deliver parcels without any human intervention. In this case, the customer receiving delivery would be notified to come out to the truck and retrieve the package. Upon arrival at their door, customers would take the parcel from the specified locker mounted on the truck, using a passcode sent to the recipient or using a Bluetooth connection to the recipient's phone.

## Autonomous Food Delivery

In contrast with package delivery, the recipient for food delivery is always there to receive the delivery. This advantage will make autonomous food delivery very compelling. For $3-$5 a restaurant can quickly make an autonomous food delivery. Autonomous order delivery vehicles can be custom designed. They can be small. Think automated scooters designed for holding food and keeping it warm. They can be simple, small and inexpensive.

Companies such as Domino's Pizza will be able to replace many of its delivery people with autonomous delivery vehicles. Many national chains such as Domino's Pizza will find it profitable to invest in a fleet of autonomous delivery vehicles. Domino's DRU (Domino's Robotic Unit) is a prototype autonomous delivery vehicle. With sleek, refined forms combined with a friendly persona and lighting to help customers identify and interact with it. The DRU claims to be the world's first of its type for autonomous commercial delivery. It's a four-wheeled vehicle with compartments built to keep the customer's order piping hot and drinks icy cold while traveling at a safe speed from the store to the customer's door. The DRU can navigate from a starting point to its destination, selecting the best path of travel. Its onboard sensors enable it to perceive obstacles along the way and avoid them if necessary. It's not clear at this point whether the DRU is intended to travel on sidewalks or roadways. There are 13,800 Domino's Pizza locations. If each had 3-4 autonomous delivery vehicles, they would have a fleet of approximately 50,000.

Mainstream restaurants will be able to add affordable delivery options. While much of this future autonomous delivery will be a service, large restaurants will own their delivery fleets. These automated delivery vehicles will also be feasible with sufficiently-autonomous capabilities. They may not be able to deliver to all addresses initially (just as Domino's doesn't deliver to all locations today), but progressively the number of destinations will increase.

The cost of delivery will drop dramatically, and convenience will increase. The estimated cost for pizza delivery is approximately $4-$5, plus a tip, or roughly $7-$8 for a $15-$18 pizza. So, the cost of buying a delivered pizza would be 30%-50% lower. Automated delivery vehicles will notify you ahead of arriving, and you will be able to track delivery along the way. The food will be kept warm and the drinks cold along the way. Companies should be able to have a more extensive fleet of autonomous vehicles than having delivery drivers on hand, so there will be more delivery capacity. On the downside, autonomous delivery vehicles won't be able to carry your food upstairs; you will need to come out to get it.

Food delivery won't just be limited to fast food either. There will be market opportunities for nice restaurants to use autonomous food delivery services to deliver custom ordered dinners to customers. Employees can also request automated delivery of individually selected lunches.

## Autonomous Shuttle Services

Shuttle services transport people a short distance, typically along a predetermined route. Airport shuttle services taking arriving passengers from the airport to rental car facilities, parking lots, or hotels are the best example. Shuttles that transport students around a college campus are another example.

While convenient, shuttle services can be expensive. Some airports use large extended length shuttles to and from car rental facilities to reduce labor costs. Hotels limit the number of trips to scheduled hours and try to use employee drivers who do other work. Labor costs are what make shuttle services expensive.

Autonomous shuttle services will be an early use of AVs. There are some significant benefits in the elimination of labor costs and increased convenience because they can pick up passengers as soon as they want, in-stead of waiting. Since shuttle services usually follow a predetermined route or small set of paths, autonomous shuttle vehicles will be easier to program and test.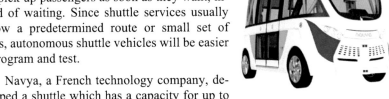

Navya, a French technology company, de-veloped a shuttle which has a capacity for up to 11 seated passengers and four standees. Maximum operating speed during its tri-als is 7 km/hr., and the vehicle can operate for up to 13 hours on a single charge. Navigation and obstacle detection uses a four-step process that combines radar, lidar, GPS and camera scanning, along with the algorithmic processing of geo-graphical and geotechnical data. The company employs a 'route scrubbing' method when the vehicles are initially deployed, allowing the primary camera and scanning data to be analyzed and refined by back-office teams before deploying the vehicles with passengers.

## Autonomous Bus Transportation

Bus transportation in major cities has been declining. For example, New York City is facing several significant issues with bus transportation. Bus rid-ership has seen a steady decrease for nearly the last two decades.

Industry professionals believe automated buses are the key to changing an outdated public transportation system. Creating a more fluid transportation system may encourage more people to start relying on buses rather than taking taxis, us-ing rideshare apps, or driving. I'm not sure that I agree with that.

AVs will create more convenient alternatives than buses. Autonomous ride services, for example, may be so inexpensive that bus transportation won't be able to compete with individual rides. Autonomous buses also will be more expensive and challenging to develop. A large bus is more difficult to react and maneuver. Smaller shuttle bus services may displace large buses.

There may be an opportunity for long-range, cross-country trips on auto-mated buses by companies like Greyhound Bus. These buses could have a "driver" on board who can sleep along the way and handle the bus for stops or other needs.

# Military AVs

The U.S. Army and other world militaries are developing fully autonomous vehicles primarily to protect the soldiers who are in the convoys transferring fuel, food, water, and other essential supplies. The military has a compelling case for moving convoys of vehicles through dangerous areas without putting soldiers at risk.

Military vehicles must be able to operate in harsh conditions, without paved roads, and under enemy fire. One plan is the "leader/follower" system for logistics. A manned lead vehicle that's followed by a series of trucks or an unmanned vehicle in the front will avoid injuries if it hits an explosive.

An example is the TerraMax vehicles made by Oshkosh Defense. It is a modular appliqué kit for any tactical wheeled vehicle. A single operator can supervise multiple unmanned vehicles while providing beyond line-of-sight situational awareness. TerraMax vehicles function autonomously across the complex terrain, day or night, in any weather conditions and seamlessly collaborate with manned vehicles.

There could be a ten-truck convoy with two humans in the lead truck and nine automated trucks following behind. Unmanned ground vehicles can supplement soldiers on the ground and possibly replace them in the future. They are suited to perform daily routine and tedious tasks precisely. They can be designed to withstand pressure and complete various tasks which could potentially harm or threaten the life of a soldier, thus significantly improving combat effectiveness on the battle-

field. Development of these vehicles is complex since they need to able to traverse various terrains and accomplish multiple military assignments. There are four types of unmanned ground vehicles:

- Small robotic building and tunnel searchers designed for short-distance reconnaissance remotely operated by a soldier. These machines are especially useful in urban areas with cramped spaces, tunnels, caves, and ruins which can potentially provide a lot of hiding places for enemy traps and other dangers. They use high-resolution cameras (infrared and night-vision), and a variety of sensors mounted with a robotic arm for specific situations.
- The small-unit logistic mover also known as the Donkey is a concept of weapons supply and even wounded soldier carrier, which could operate

across the battlefield. This small-unit logistics mover is a semi-autonomous medium-size vehicle that precedes or follows with the capacity to carry a couple of hundred kilos.

- The unmanned wingman ground vehicle is a medium-sized unmanned vehicle which aids small mechanized infantry and armor units that usually operate in small teams. Assistance is provided by constant surveillance and observation of the battlefield thus providing crucial information to the combat units. With high-tech cameras, sensors, and mobility the "Wingman" can do non-stop scout/guard duty.
- An autonomous hunter-killer team vehicle concept is based on the idea that multiple unmanned vehicles can perform tasks such as ambush and termination missions while penetrating deep into the frontline of the battlefield in various terrains and weather conditions. Hunter-killer teams could operate day and night, in all terrain/weather conditions, armed with sophisticated sensors, lethal weapons and with a high-degree of autonomy. These vehicles will be capable of high autonomy, meaning that they are not dependent on the human operator.

The debate about fully autonomous weapons is intensifying. Lawyers, ethicists, military personnel, human rights advocates, scientists, and diplomats argue about the legality and desirability of weapons that would select and engage targets without meaningful human control. Divergent views remain as military technology moves toward ever greater autonomy, but there are mounting expressions of concern about how these weapons could revolutionize warfare as we know it. Replacing human troops with machines could make the decision to go to war much more comfortable; it's merely a matter of dispatching some equipment with no military risk. I tend to agree with the need to restrict such weapons.

# United States Postal Service

The United States postal service is preparing for a future where drivers won't need to do all the driving to deliver mail. It has been developing a self-driving mail truck and identified five ways that an autonomous vehicle could improve mail delivery.

## Suburban and Rural Routes

The most likely case is where the driver doesn't need to leave the truck. In suburban and rural routes, in which mail carriers never leave the mail truck, couriers will spend their time organizing and sorting mail while the truck handles the driving duties. This process could be more time-efficient, reducing or even eliminating the need for additional stops to sort future deliveries.

## Inner City Routes

Based on a German governmental white paper, a self-driving mail truck's ability to hunt for its parking spot will save upwards of 40 minutes each day. The truck would drop off the courier along the designated route. The courier would then deliver the mail as usual, while the truck heads off in search of a place to park. When finished, the courier would merely summon the truck via an app, and continue to the next route.

### Strip Malls and Some Suburban Routes

In some locations like strip malls or some suburban locales, the vehicle's
ability to navigate around com-
plex obstacles — parked cars and
pedestrians, among others —
would enable it to follow the courier rather than have the courier
drive, thus reducing the walking
distance required to deliver heavy
packages. The most tangible benefits come in the form of a less
fatigued carrier, but the USPS
considers the time saved walking
back to the vehicle to be significant.

### Replenishment of Mail

Even if delivery routes themselves were to remain unchanged, self-driving
mail trucks would cut down on wasted time by eliminating the need for couriers
to return to the post office after delivering a truck's mail load. A second truck
could re-stock the courier with a fresh batch of parcels.

### Mobile P.O. Boxes

In the more extreme and further-out case, a self-driving vehicle could function as a mobile P.O. Box. Once the customer delivers the package to the post
office, they can schedule a delivery at any time of the day, meet the vehicle
curbside, enter a code to access its locked compartment, and retrieve the package.

# Chapter 6
# Technologies Enabling
# Autonomous Vehicles

Autonomous driving is enabled by some exciting new technologies. In chapter 2, I provided a functional overview of *what* autonomous vehicles (AVs) do to drive at different levels of autonomy. In this chapter, I'll explain *how* AVs work and describe the new emerging technologies that make it possible.

A significant amount of money has been invested in the technologies that make AVs possible. I estimate that more than $100 billion has already been invested or committed to the development of AVs. This includes the R&D done by car manufacturers and the acquisitions they have made. As an example, based on a GM investor presentation it plans to have 2,100 people working on AVs in 2018, which I estimate equates to a $400-$500 million annual investment. The overall investment in AVs includes the R&D investment that is not disclosed by major technology companies, specifically Apple and Google (Waymo). Based on a legal disclosure, Google invested more than $1 billion through 2015 alone. The overall investment also includes the R&D investment by numerous companies developing the sensors, processors, and detailed maps, many of which are start-up companies with venture capital funding. Intel acquired Mobileye for $15 billion.

As you will see, some of the technologies already developed will become more affordable. Others are in the early stages of development, but they are proven to be feasible and just need to be completed and brought to market. In some cases, there is not yet universal agreement about which technologies to use to solve specific problems, which is typical of most technological revolutions. The important point is that all the technologies to make AVs are feasible.

An autonomous vehicle requires a variety of sensors to "see" and very quickly interpret its environment. The primary sensors use camera-based vision, lidar (Light Detection and Ranging), sonar, and radar technologies. The output from these sensors is processed and interpreted instantly by a high-powered microprocessor-based computer system onboard the car, which is operated by sophisticated software using artificial intelligence. The computer system using detailed maps initiates the actions to "drive" the vehicle by braking, accelerating, and turning the vehicle.

## Sensors

Sensors are essential to autonomous vehicles. They instantly enable AVs to sense and identify the characteristics of their surroundings. There are different configurations of sensors, depending on the sensor design strategy of the AV manufacturer, but the following diagram from Texas Instruments is a typical configuration.

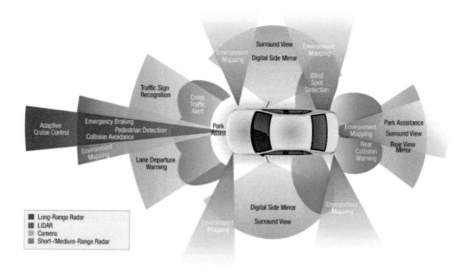

This sensor diagram illustrates how the different types of sensors perform various functions. Long-range radar provides adaptive cruise control. Short-range radar provides cross-traffic alert and rear collision warning. Cameras provide traffic signal recognition and surround view. Lidar provides mapping of the environment.

Also, you can see in the illustration, that they overlap with each other. There are trade-offs among these sensor technologies. Cameras are effectively the autonomous replacement for a driver's eyes, letting the car 'see' what's happening in the world around it, but they provide a full surround vision that drivers don't have. Radar can accurately measure an object's motion and speed. Lidar is the most powerful sensor. When it comes to cost, cameras are cheaper than radar or lidar, but the latest high-definition cameras require powerful processors to digitize and interpret the millions of pixels in every frame.

The safest approach is to use an array of sensors to build in redundancy and offset the limitations of any one type of sensor. Each type of sensor solves a different part of the sensing challenge, and sensor-fusion software integrates all of this varied input.

Here is another example using an array of sensors in a General Motors Cruise system for its AVs. It includes 14 cameras, 3 articulating radars, 5 lidars, 8 radars, and 10 ultra-short radars.

## AV SPECIFIC REDUNDANT HARDWARE SYSTEMS

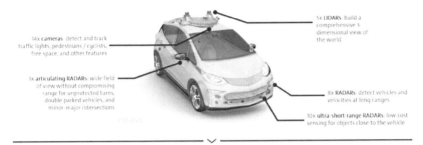

5x LIDARs: build a comprehensive 3-dimensional view of the world

14x cameras: detect and track traffic lights, pedestrians / cyclists, free space, and other features

3x articulating RADARs: wide field of view without compromising range for unprotected turns, double parked vehicles, and minor-major intersections

8x RADARs: detect vehicles and velocities at long ranges

10x ultra-short-range RADARs: low-cost sensing for objects close to the vehicle

The cameras detect and track stationary and moving objects such as traffic lights and pedestrians. The articulating radar provides a wider field of view to detect turns, double-parked vehicles, and intersections. Lidar is used to provide the detailed 3D mapping. The non-articulating radar detects vehicles and their velocities at a longer range. The ultra-short-range radar is a low-cost sensor for sensing objects close to the vehicle.

Let's look at the technology in each of these sensors.

### Cameras

Video cameras are the primary sensors for autonomous vehicles. Cameras excel at classification and texture interpretation, and they are the cheapest and most available sensor. However, cameras use massive amounts of data (full HD means millions of pixels in every frame), making processing video input computationally intense and algorithmically complex. Unlike both lidar and radar, cameras can see color, making them the more useful for scene interpretation. Cameras are the cheapest sensor of the three, and they will likely remain important for the foreseeable future. However, the use cases will be dependent on the software algorithms and processing power for the massive amount of data generated.

There are also different varieties of cameras for various functions: narrow-focus cameras for longer distance, wider-focus cameras for shorter range, side-facing cameras, etc. Cameras can capture color or monochrome images. Color images can best interpret objects since color can be meaningful (traffic lights are the best example). Monochrome images require less processing power than color; so, there are some trade-offs.

Tesla has chosen to use cameras instead of lidar. This approach is much cheaper, and the cameras can be built into the car to make them almost invisible. As illustrated, Tesla uses eight surround cameras to provide 360-degrees of visibility around the car at up to 250 meters. Twelve ultrasonic sensors complement this vision system, allowing for detection of both hard and soft objects at nearly twice the distance of the previous Tesla system. A forward-facing radar with enhanced processing provides additional data on a redundant wavelength that can see through heavy rain, fog, dust and even the car ahead because cameras struggle to see in these conditions.

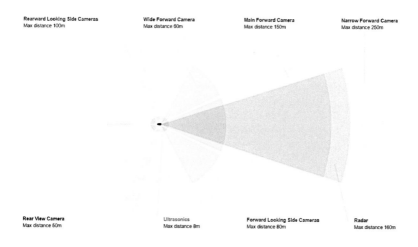

Tesla uses monochrome cameras instead of color in its autopilot system to avoid overloading the car's processor, but color helps in object recognition. Color images can be semantically segmented, such as identifying street lamps, trees, and buildings. Color allows AVs to interpret better what they see. For example, color cameras can interpret a bright white truck in front of a bright blue sky. This limitation of monochrome cameras was the cause of a fatal full-speed accident in 2016 involving a Tesla, which I want to clarify was blamed on the driver and other circumstances, and not the car's autopilot.

> *"Neither the autopilot nor the driver noticed the white side of the tractor-trailer against the brightly lit sky, so the brake was not applied,"* Tesla wrote on June 30, 2016. *"The car was headed east on a sunny afternoon, and the driver was not paying attention. The car's forward-looking monochrome camera couldn't distinguish the color difference between the truck and the sky. The car's forward-looking radar was ignored because it could mistake overhead road signs for vehicles".*

Cameras will be a primary sensor for all autonomous vehicles. The open questions are: (1) can they be used exclusively instead of lidar to reduce costs, (2) will monochrome versions work sufficiently to minimize processing requirements or will color be necessary, and (3) how much processing power will be required to processes video images instantaneously?

## Radar

Radar is an object-detection system that uses radio waves to determine the range, angle, and velocity of objects. A radar system consists of a transmitter producing electromagnetic waves in the radio or microwave spectrum, a transmitting antenna, a receiving antenna (often the same antenna is used for transmitting and receiving), and a receiver and processor to determine properties of objects. Radio waves (pulsed or continuous) from the transmitter reflect off the object and then return to the receiver, giving information about the object's location and speed.

Radar sensors can be classified by their operating distance ranges: short-range radar (SRR) 0.2-30-meter range, medium-range radar (MRR) in the 30-80-meter range and long-range radar (LRR) from 80 meters to more than 200 meters' range. Long-range radar is the sensor used in adaptive cruise control and highway automatic braking systems, which is illustrated in this Audi cruise control image.

Radar is excellent for motion measurement. It requires less processing power than cameras and uses far less data than a lidar. While less angularly accurate than lidar, radar can work in every condition and even use reflection to see behind obstacles.

LRR radar sensors sometimes have limitations. They might not react correctly to certain conditions such as a car cutting in front of a vehicle, thin profile vehicles such as motorcycles being staggered in a lane, and determining distance based on the wrong vehicle due to the curvature of the road. Radar also has lower resolution. To overcome the limitations in these situations, a radar sensor could be paired with a camera sensor to provide additional context to the detection. Many AVs use a combination of radar and lidar to "cross-validate" what they're seeing and to predict motion.

Even though radar may often pick up the same surrounding objects as other sensors, it doesn't need line of sight because radio waves can be reflected, which makes it more efficient than cameras and lidar in poor weather conditions.

The FCC expanded the frequency allocations for vehicle radar. Initially limited to 76-77 GHz, the FCC extended the vehicle radar allocation to 76-81 GHz, taking over frequencies allocated to radiolocation and amateur radio operators.

## Lidar

Lidar (an acronym for Light Detection and Ranging) is an active remote sensor that works on a similar principle to radar but uses light waves instead of radio waves. Lidar was initially invented in the 1960s, just after the development of the laser. The Apollo 15 mission in 1971 used lidar to map the surface of the moon, giving everyone the first glimpse of what it could do.

Lidar is the most powerful of the three sensors, and, unsurprisingly, it's also by far the most expensive. Current lidar systems use 64 beams and sit on the roof of the vehicle, identifying all detail to the centimeter from 100 meters away in all directions. It can generate a comprehensive and precise high-definition 3D map of the immediate world around it. Unlike camera-generated images that require pattern recognition software to make the data useful to onboard computers, lidar systems create a stream of data-point clouds that are immediately useful to computers. Lidar systems can render one million data points per second covering a 360-degree field of view with a range of 393 feet through low-visibility conditions. Both the level of detail and distance are critical to autonomous driving, especially at higher speeds. At 90 miles per hour, stopping distance is approximately 400 feet, so detecting an object that far out can be critical.

Lidar is a series of rotating, stacked lasers that shoot out at different angles. Two laser beams form what is called a channel. The signal from each channel creates one contour line and put together those lines generate a 3-D image of the surrounding environment. The more lasers in each stack, the higher the resolution. Velodyne, for instance, manufactures lidar products with 16, 32 and 64 laser channels.

Fundamentally, a lidar sensor illuminates a target area with invisible light and measures the time it takes for this light to bounce back to its source. As light moves at a constant speed, it calculates the distance between the sensor and objects with high accuracy.

Velodyne LIDAR

Today, most, but not all, experts agree that lidar is one of the critical sensing technologies required to enable fully and sufficiently autonomous driving. There is also general agreement that lidar technology is currently too expensive, and not aesthetically acceptable as a big ugly system sitting on top of the vehicle's roof.

Solid-state versions of lidar are currently being developed. They are hardly visible in cars and are cheaper. Conversely, solid-state lidars, which are sold in higher volumes, are limited in scope to less than 90-degree optical scans and must be embedded in the front and back bumpers of cars to achieve a comprehensive view of the surroundings. In this way, there are four devices integrated into the car with two on each bumper. Solid-state devices also require extra software to manage the four separate data streams from the four devices.

This chart illustrates GM's expectations of the cost reduction of lidar from currently available at about $20,000 to next generation at half of that cost to an eventual cost of about $250 with twice the range.

| Currently Available LIDAR | • Effective range: 1x<br>• Cost: ~$20,000<br>• Quality Issues |
| Next Gen LIDAR | • Expected effective range: ~1.25x<br>• Cost: ~$10,000 |
| Strobe + GM + Cruise | • Expected effective range: ~2.5x<br>• Cost: ~$300 |

The cost and attractiveness of lidar may not matter as much initially because they are expected to be deployed in limited fleets of autonomous ride service vehicles, which can afford the investment. And as a rider, when you hail a car as part of an autonomous ride service, you don't care that much if there is a large lidar device on top of the car.

Waymo is focused on lidar as its primary distinguishing technology. CEO John Krafcik stated that Waymo built the entire sensor suite used by its self-driving Chrysler Pacifica Hybrid test vehicles. This was a significant accomplishment for the company, making Waymo no longer vulnerable to third-party suppliers. According to a Bloomberg article, the company worked on scalability, leading to a 90% decrease in the cost of the lidar sensor. Krafcik also told Bloomberg the new sensor package on the Waymo Chrysler Pacifica is "highly effective in rain, fog, and snow," which have typically been trouble for lidar systems because of the reflective nature of water in the air.

Waymo later accused Uber of stealing its work, claiming a former Google employee downloaded 14,000 technical files from a company server, then used the information to launch the autonomous truck startup, Otto. Uber acquired Otto a few months later and appointed him to lead its AV program. This lawsuit was recently settled in Waymo's favor.

Lidar also has another important use. It can create detailed high-definition digital maps for autonomous driving (this will be discussed later). On mapping vehicles, the lidar units are usually mounted high, to get the necessary height for modeling, and tilted down. As the vehicle drives along the road, the lidar units capture a 3-D model of the road and its surroundings. It effectively builds the model by taking slices of the outside environment as the vehicles travels.

## Ultrasonic Sensors

Ultrasonic proximity detectors measure the distance to nearby objects using sensors located in the front and/or rear bumper or visually minimized within grills or recesses. The sensors emit acoustic pulses, with a control unit measuring the return interval of each reflected signal to calculate object distances. In a semi-autonomous vehicle, the system can warn the driver with acoustic tones or visual warnings, such as LED or LCD readouts to indicate object distance. In an autonomous vehicle, ultrasonic sensors serve multiple purposes. A Tesla, for example, has 12 ultrasonic sensors providing the system with 360-degree vision at a range of about 5 meters. They are used by the auto parking system to map out parking spots and by the autopilot for changing lanes into traffic.

The theory behind the ultrasound sensor is based on echolocation (like SONAR, the same thing bats use to navigate). The frequency of the sound is so high that humans cannot detect it, so it is inconspicuous. As sound hits a solid object, it is reflected creating an echo. Since the speed of sound is known and constant for similar conditions (such as wind or humidity), it is possible to determine the distance of the object from an echo by multiplying the speed of sound by half the time it takes to hear the echo.

As an ultrasonic system relies on the reflection of sound waves, the system may not detect flat objects or object insufficiently large to reflect sound, for example, a thin pole or a longitudinal object pointed directly at the vehicle or near an object. Objects with flat surfaces angled from the vertical may deflect return sound waves away from the sensors, hindering detection.

## Electronic Control Systems

In autonomous vehicles, electronic control systems are embedded systems that control one or more of the electrical systems or subsystems, such as the throttle, brakes, steering, and braking. This is an obviously needed change from systems that required a human to shift gears, step on the brakes or gas, etc.

### Electronic Throttle Control

The electronic throttle control (ETC) system electronically connects the accelerator pedal to the throttle, replacing mechanical linkage. The system uses software to determine the required throttle position by calculations from data measured by other sensors, including the accelerator pedal position sensors, engine speed sensors, vehicle speed sensors, and cruise control switches. In AVs, the electronic control system eliminates the accelerator pedal position and uses the control signals from the primary computer system instead.

Although ETC is an enabling technology for AVs, it has been introduced in many cars already. It makes vehicle power-train characteristics seamlessly consistent irrespective of prevailing conditions, such as engine temperature, altitude, and accessory loads. It also improves the ease with which the driver can execute gear changes and deals with the torque changes associated with rapid accelerations and decelerations. ETC facilitates the integration of features such as cruise control, traction control, stability control, and other functions that require torque management since the throttle can be moved irrespective of the position of the driver's accelerator pedal. ETC also provides some benefit in areas such as air-fuel ratio control, exhaust emissions and fuel consumption reduction, and it also works in concert with other technologies such as gasoline direct injection.

## Brake-by-Wire

Drive-by-wire technology replaces the conventional mechanical and hydraulic control systems with electronic control systems using electromechanical actuators and human-machine interfaces.

Brake-by-wire is a term frequently used to describe the technology that controls brakes through an electrical system instead of a mechanical one. It can supplement conventional brakes, or it can be a standalone brake system, as it will be in AVs. The technology replaces traditional pumps, hoses, fluids, belts, vacuum servos and master cylinders with electronic sensors and actuators.

Brake-by-wire technology has been widely commercialized with the introduction of electric and hybrid vehicles.

## Steer-by-Wire

Originally steering used a rack and pinion system to direct a car in the desired direction. Next came hydraulically-assisted steering, which became popular starting in the 1950s. Then hydraulics were replaced with electric motors. The motors are usually placed either at the base of the steering column or directly on the steering rack.

Steer-by-wire is frequently used to describe an electronic system used to steer a vehicle automatically. It replaces the traditional steering control of a car that uses components and linkages between the steering wheel and the front wheels. The control of the wheels' direction is established through electric motors which are actuated by electronic control units monitoring the steering wheel inputs from the driver, or in the case of AVs, from a central computer system.

Unlike the two previous systems, automatic steering control is relatively new and not yet approved as an independent control system for many uses. Even in a Tesla, the steering wheel is still mechanically connected to the front wheel, although if the promise of autopilot is real, then the steering system must be capable of operating automatically without any driver control. It will be one of the new requirements needed for AVs.

## Shift-by-Wire

Shift-by-wire is a term typically used to describe the system by which the transmission modes are engaged and changed in a vehicle through electronic controls without any mechanical linkage between the gear shifting lever and the transmission. Transmission shifting was traditionally accomplished by mechanical links to put the vehicle in Park, Reverse, Neutral and Drive positions through a lever mounted on the steering column or a gear shifter near the center console. A shift-by-wire transmission system enables a driver to manually select the next desired gear, and the system completes the shift automatically.

In more vehicles, mechanical linkages are disappearing in favor of shift-by-wire. Yet, in some semi-autonomous vehicles, such as Mercedes, the driver still needs to change gears for automatic parking. Shift-by-wire will be necessary for all AVs to shift gears.

## Computer Processing

AVs will require enormous computing power, which is one of the reasons they haven't been feasible until now. Approximately a gigabyte (equal to roughly a billion bytes) of data must be processed each second by an AV's real-time operating system. This data will need to be analyzed instantly so the vehicle can react in less than a second. For example, it will need to figure out when, how hard, and how fast to brake based on analysis of a range of variables, from the vehicle's speed to the road conditions to surrounding traffic. It will need to successfully gauge the flow of traffic to merge onto a freeway and account for the unpredictable behavior of pedestrians, bicyclists, and other cars while in the city.

AVs require a new computer architecture. The primary processing functions will be more centralized. Today a typical car can have between 25 and 50 processors, controlling blind-spot and pedestrian collision warnings, automated breaking and maintaining a safe-distance via smart cruise control, and many others. Many of these functions have their own computer software. This has spawned a distributed-computing approach that accommodates this growing ecosystem of embedded control units. But with each new addition, complexity and cost increase, as do the challenges of integrating so many disparate systems. There are many benefits in a more centralized processing model for AVs.

The companies developing AVs disagree on the computing capacity needed to achieve autonomous driving, and this disagreement results in as many different computing solutions as there are autonomous vehicle programs. Tesla reached level 2 with 0.256 trillion of operations per seconds (TOPS) in Autopilot 1.0, but level 3 to 5 are expected to need significantly greater capacity – anything from 2 to 20 TOPS. Nvidia's new processing platform provides 310 TOPS, which it believes is the requirement for autonomous driving. In my opinion, the highest level of computing power will be required.

However, while the processors in AVs must deliver increasing computing power, they also must do so as efficiently as possible, which means managing the amount of power required, as well as cooling.

Several computing companies are developing specialized processors for AVs. They combine experience in powerful processing, graphics processing, and complete systems. One of these, Nvidia, has created a lot of early momentum and provides a glimpse into the processing requirements with its Drive PX platform, particularly its Pegasus system. The Drive PX Pegasus computer platform was designed to control AVs. The platform must not only have sufficient computing power to process the data of an entire array of sensors in real time and derive driving decisions from it, but it also must meet the highest requirements for functional safety. To be resilient and fail-safe, it is equipped with multiple different processors that serve as a back-up for each other.

To achieve this computing power, the PX Pegasus is equipped with four processors - two SoCs (literally systems on a chip) and two next-generation GPUs, each designed specifically for deep learning and autonomous driving. The Pegasus has computing power of 320 TOPS (this means trillions of operations per second), ten times as much as Nvidia's previous AI platform Drive PX 2. The platform is designed for updates via the over-the-air interface (OTA) to keep the software running on the computer up-to-date at all times. The AI computer is due to be delivered in the second half of 2018; a collection of development tools and libraries is already available under the name Drive Works.

*Nvidia's PX Pegasus has computing power of 320 trillion operations per second.*

The high demands on reliability and redundancy mean that these computers require very high computing power. Compared to semi-autonomous vehicles, sufficiently-autonomous and fully-autonomous vehicles need overlapping surveillance. The vehicle itself must continuously know its position to the centimeter and immediately recognize other vehicles and people in the vicinity. Because of these requirements, sufficiently and fully autonomous cars require 50 to 100 times more computing power semi-autonomous vehicles.

## Software

Software provides the intelligence for autonomous vehicles. AV software is extensive and complex, and it requires a tremendous investment. Most of the companies developing AVs are developing their own software because this will be the defining technology of their AV platforms. There are also some smaller companies specializing in developing AV software. Autonomous driving demands some of the most complex, real-time computing capabilities ever developed. The good thing about software is that while it has a high development cost, the cost to have it installed in each vehicle is negligible.

*Software is the defining technology of AV platforms.*

### AV Software Architecture

The architecture, or overall structure of AV software, defines the different functions and how they work together. This architecture is critical for the development and maintenance of this sophisticated software. Major companies developing AV software keep their architecture proprietary since it provides a competitive advantage. So, to illustrate what AV software architecture looks like, I'll use an example from one of the successful entrants in the DARPA Urban Challenge. This is the software architecture used by Team Victor Tango's Odin from Virginia Polytechnic Institute and others. It was developed using the LabVIEW software platform from National Instruments.

At the left side of this architecture is the data coming from laser (lidar) and camera sensors, as well as GPS data. The first set of software modules address *Perception*. The *Localization* module takes input data from the sensors and GPS and uses perception software algorithms to position the vehicle where it is on the road. Vehicle localization requires accuracy and 6-degrees of freedom. This is the freedom of movement of a rigid body in three-dimensional space. Specifically, the body is free to change position as forward/backward, up/down, left/right translation in three perpendicular axes.

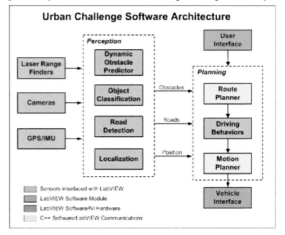

*Road Detection* localizes the road and identifies markings on the road. The *Object Classification* module uses sensor data to detect and interpret the characteristics of other objects, such as vehicles and pedestrians. There are two types of objects: static (fixed and unmovable) and dynamic (capable of and expected to move). The *Predict* module handles dynamic objects and predicts where the object will be in a second, few seconds, etc. The output of the Perception set of modules provides the vehicle position, roads, and objects to the *Planning* set of modules.

The *Planning* modules plan the movement of the AV. The *Route Planner* module takes the vehicle location data (where it is) and adds the data on where it wants to go, then it plans the route it needs to take to get there. For example, proceed 50 meters in the right lane and then make a right turn into the correct lane. The *Driving Behaviors* module adjusts the timing and possibly the planned movement based on how other objects might move. For example, there is a car ahead, and the AV needs to wait for that car to proceed. From that data, the *Motion Planner* determines the specific operations the car needs to follow.

Finally, the *Vehicle Interface* software ties into the steering, throttle, brake, etc. It sends the appropriate instructions to these systems to initiate and continue the specific maneuvers.

## Sensor Fusion

In simple terms, sensor fusion is the stitching together of data from various sensors, typically video, radar, lidar, ultrasonic. Sensor fusion software captures raw images and allows an AV to build a 360-degree model of its surroundings instantly.

Sensor fusion does much more than just reading sensor data. Each sensor type has shortcomings that cannot be overcome using the same sensor. For example, a camera working in the visible spectrum has trouble in dense fog, rain, sun glare and the absence of light. Radar lacks the high-resolution of lidar sensors, but it is excellent for measuring distances through rain and fog. Cameras don't work well in these conditions or the absence of light, but they can see color and have a high resolution.

Sensor fusion takes the inputs of different sensors and sensor types and uses the combined information to perceive the environment more accurately. That results in a better and safer decision than independent systems could make.

An example of sensor fusion is combining the functions of a front camera with the front radar. The front radar can measure the speed and distance of objects up to 150 meters in all weather conditions, while the camera is excellent in detecting and differentiating objects, which includes reading street signs and street markings. By using multiple camera sensors with a different field of view and different optics, such things as pedestrians and bikes passing in front the car, as well as objects 150 meters and more ahead, can be identified.

Using different sensor types also offers a certain level of redundancy to environmental conditions that could make all sensors of one kind fail. Such a sensor-fused system could maintain some basic or emergency functionality, even if it lost a sensor.

Although the core concept underlying sensor fusion software algorithms seems simple, its implementation calls for a variety of development tools and a broad spectrum of skill sets. The design community requires software development platforms that enable the creation of sophisticated and customized algorithms to accommodate the unique requirements of these applications. Sensor fusion software must be capable of a high level of abstraction to support flexible integration, different kinds of sensors, and a variety of algorithms. Developers must be careful to avoid data overload and latency – most sensor fusion processing must be done in real time. And the algorithms themselves must be able to accurately react to every possible outcome of human behavior (including irrational behavior).

There are many different types of sensing technologies, but the critical issue is the fusion of these technologies into an integrated platform that enables the vehicle to accurately assess the world around it and drive accordingly.

## Simultaneous Localization and Mapping

Simultaneous localization and mapping (SLAM) is a technology process where an AV can create a map of its surroundings, and orient itself correctly within this map in real time. This is no easy task, and it currently exists at the frontiers of technology research and design. A significant roadblock to successfully implementing SLAM technology is the chicken-and-egg problem introduced by the two required tasks. To successfully map an environment, the vehicle must know its orientation and position within that environment; however, this information comes from a pre-existing map of the environment.

SLAM technology typically overcomes this complex chicken-and-egg issue by building a pre-existing map of an environment using GPS data. This map is then iteratively refined as the AV moves through the environment. The real challenge of this technology is one of accuracy. Measurements must continuously be taken as the AV moves through space, and the technology must consider the "noise" that is introduced by both the movement of the AV and the inaccuracy of the measurement method. This makes SLAM technology primarily a matter of measurement and mathematics.

As an example of this measurement and mathematics in action, one can look at the implementation of Google's AV. The AV primarily takes measurements using the roof mounted lidar, which can create a 3D map of its surroundings up to 10 times a second. This frequency of evaluations is critical as the AV moves. These measurements are used to augment the pre-existing GPS maps, which Google is well known for maintaining as part of its Google Maps service. The readings create a massive amount of data, and generating meaning from this data to make driving decisions is the work of statistics. The software on the car uses some advanced statistical algorithms, including Monte Carlo models and Bayesian filters to accurately map the environment.

### Simulation

Simulation software is used to educate and train autonomous vehicles. It simulates "real world" situations that a vehicle will encounter, tests its response, and then modifies it as necessary. Simulation software primarily runs outside of the vehicle in development labs. What it learns from the simulation is then transferred to the self-driving software on each vehicle.

There are several advantages to simulation. First, a simulation can "drive" a lot more miles than would be possible to do physically. It can do millions of miles per day. Imagine a new driver having to simulate a million miles of driving before getting a license. Second, simulations can focus on the exciting and challenging situations drivers encounter, rather than the standard boring mile driven. Third, using simulation the development time for the software can be much, much faster. And fourth, all the simulated situations can be saved and then used to test future releases of the AV software.

All companies developing autonomous systems use autonomous driving simulation software, but they tend to keep the efforts secret. Waymo first discussed its autonomous simulation software with Atlantic Magazine for an article by Alexis Madrigal on August 23, 2017. The software, called Carcraft (named after the software game World of Warcraft), simulates eight million miles of virtual driving per day. Here are some of the highlights:

- At any time, 25,000 virtual self-driving cars are making their way through fully modeled versions of Austin, Mountain View, and Phoenix, as well as test-track scenarios. Waymo might simulate driving down a particularly tricky road hundreds of thousands of times in a single day. In 2016, the simulations logged 2.5 billion virtual miles compared to a little over 3 million miles by Waymo's real-life self-driving cars that run on public roads. And crucially, the virtual miles focus on what Waymo people call "interesting" miles where they might learn something new. These are not dull highway commuter miles.

- The software has a model of the city of Phoenix. It shows where all the lanes are, which lanes lead into other lanes, where stop signs are, where traffic lights are, where curbs are, where the center of the lane is, etc. Basically, it is everything you need to know. They zoom in on a single four-way stop some-  where near Phoenix and start dropping in synthetic cars and pe- destrians and cyclists. Then the objects on the screen begin to move. Cars act like cars, driv- ing in their lanes and turning. Cyclists act like cyclists. Their logic is modeled from the millions of miles of public-road driving Waymo has done. Underneath it all, there is that hyper-detailed map and models for the physics of the different objects in the scene. Iin this case, there are 800 scenarios generated by this four-way stop.

- Not surprisingly, the hardest thing to simulate is the behavior of the other people. It's like the old parental concern: "I'm not worried about you driving. I'm worried about the other people on the road."

- There is one fundamental difference between this and the real world. In the real world, they must take in fresh, real-time data about the environment and convert it into an understanding of the scene, but for most simulations, simulations skip that object-recognition step. Instead of feeding the car raw data it has to identify as a pedestrian, they merely tell the car: a pedestrian is here.

## Machine Learning

It's easy to envision AV software as large programs consisting of millions of elaborate rules specifying how the car should act for every conceivable situation. The typical example is an If/Then case: "If a ball rolls across the road then reduce the speed and watch for children that might come running after the ball." This example implies that there is a theory of driving implemented in the software, which there is not. It examines the images and guesses the object in each image. Initially, most of its guesses will be wrong, but the algorithm modifies internal parameters or parts of its structure somewhat and tries again. This process continues, discarding changes that reduce the algorithm's accuracy, keeping modifications that increase the accuracy, until it correctly classifies all images. This is called machine learning. Machine learning is the science of getting computers to act without being explicitly programmed.

Then when given a new object to interpret, the software algorithm it developed will classify it with high accuracy. For example, it will recognize someone on a bicycle as that object and then associate the characteristics it learned about how a bicycle might move.

Machine learning can also be used for actions and evaluations. Instead of supplying the vehicle with a fixed evaluation scheme from which the right move for each situation can be deduced, the programmers feed the machine learning software with many traffic situations and specify the correct action for each situation. The program then searches by itself for the best configuration of internal parameters and internal decision logic which allow it to act correctly in these cases. Just like with human drivers, it is difficult to explain why the AV exhibits a specific behavior in a new situation. Explicit rules are not specified. The behavior decision results from the many traffic situations to which the algorithm had been exposed beforehand. That's why AV software developers are "training" the machine learning systems with millions, probably billions, of cases through simulation. Waymo has driven more than two million miles on public roads with test drivers and has assembled an enormous source of traffic situations from which its vehicles can learn. It then supplements this with billions of simulation cases.

Another characteristic of AV software is its use of probabilistic outcomes. It maintains a distribution of outcomes with specific probabilities and decides with the highest likelihood as its current position.

Machine learning algorithms are classified into supervised algorithms and unsupervised algorithms. The difference between the two is how they learn. Supervised algorithms learn using a training data-set, and keep on learning until the algorithms reach the desired level of confidence. Unsupervised algorithms try to make sense of the available data. That means an algorithm develops a relationship within the available dataset to identify patterns or divides the data set into subgroups based on the level of similarity between them.

Apple has filed a patent application for an autonomous navigation system which learns from human driving routes. Instead of using algorithms and sensors for an autonomous system to learn the route entirely on its own, the Apple system first gets trained by driving the vehicle along a particular route. As the vehicle drives along the route, Apple's sensors develop a characterization of that route. The system progressively updates the route as it repeatedly drives the route. Once certain confidence thresholds are reached, then the autonomous navigation will turn on.

## HD Mapping (Localization)

Early developers hoped that autonomous vehicles might be able to position themselves sufficiently using the standard definition maps like the ones most people use today, along with GPS-based turning. Sensors would do the rest. That hasn't adequately worked, and now most experts realize that more precision and accuracy is required.

To solve this problem, highly-detailed, three-dimensional, computerized maps, which pinpoint a car's location and understand its surroundings, appear to be necessary. Fundamentally, the problem is that the vehicle needs to locate itself not only on the correct road, but also in the proper lane for a turn, and even how close it is to the curb or side of the roadway. These highly detailed maps, called HD (high definition) maps, provide localization. They identify precisely (how many feet or inches) where the vehicle is and where it needs to be. I say "appears to be necessary" because some experts still believe that GPS accuracy can be improved sufficiently, and these detailed HD maps won't be necessary.

Navigational maps typically locate a car's position within several yards. HD digital maps for autonomous cars must know the locations of corners, curbs and other objects within about four inches. For example, HD maps will tell the AV that it is four inches from the curb, confirm its precise location based on distance from other objects such as signs, buildings, and even trees. Then it will tell the AV that its upcoming turn takes place in six feet at the corner of the curb.

The illustration below from Here Technologies illustrates the detailed measurements from HD mapping.

Creating HD maps is a massive task. There are more than four million miles of roads in the United States, and compared with the maps used by GPS and navigation systems, the level of precision must be much higher. Waymo is creating maps for roads around its headquarters in Mountain View, Calif., and a handful of other cities, including Austin, Tex., and Kirkland, Wash. Waymo creates the maps using cars equipped with lidar units mounted on their roofs, creating images of the road and the surroundings. Engineers review the images and tag the objects that are found, like stop signs, buildings, stoplights, Etc. The laser equipment needed to do this scanning is expensive. It can cost $100,000 or more to outfit just one vehicle to do this job.

Another company, Here, which was started by Nokia and then acquired by the German automakers BMW, Daimler and Audi, is also mapping roads in the United States and Europe. It is drawing on data scooped up by scanning systems that trucking companies have agreed to install on their vehicles. Here also has its fleet of cars collecting images, and is working on algorithms to enable computers to annotate the maps.

Uber has announced a significant expansion of its internal mapping program. The Financial Times reports that the amount of the investment will be $500 million.

HD mapping is done city-by-city or metropolitan area, so AVs, particularly ARS vehicles can function in these locations once they are mapped. The chart below from GM illustrates a typical mapping process.

## HD MAPPING AND ROUTING

| Collecting | Labeling | Maintaining |
|---|---|---|
|  |  |  |
| Utilize proprietary mapping vehicle | Combination of automated and manual tools to ensure high accuracy labels | AVs update the map in real time |
| Accuracy within 3cm | Completed in cloud enabling scale | Routes adjusted based on AV data |
| Compiled using LIDAR scans, camera images, cellular coverage maps, and other sources | Overlays traffic signals in 3D space and lane markers | Accepts other data such as traffic, accident locations, construction, etc. |

*A NEW CITY CAN BE MAPPED IN ~6 MONTHS*

The first step is using lidar to collect all of the data around the vehicle with an accuracy of 3 cm. All of the objects in this data are then labeled using advanced automated tools, as well as human judgment. Once this is all done, then it needs to be continuously maintained with any changes. GM estimates that it can do the detailed maps of a new city in approximately six months.

The real world is changing all the time. HERE currently makes around 2.7 million changes to its global map database every day, illustrating just how rapid and numerous the differences are.

There is a lingering question in the industry about how useful these costly maps will be in the future. Some argue the cars will eventually get smart enough, with deep learning technology, and not need to rely on extensive maps.

China is using HD mapping to create roadblocks for U.S. automakers and tech companies to bringing AVs to the world's largest auto market. Citing national security concerns, China is limiting the amount of mapping that can be done by foreign companies. Global carmakers already need to form a partnership with a local company to open factories in China, but some are skeptical they will be able to operate their AV software in China because of the mapping restrictions.

## Vehicle-to-X Communications

Vehicle-to-X communications is a term used describe the connected vehicle, where X can be another vehicle or infrastructure such as traffic lights and street signs. Connected vehicle technology leverages advancements in wireless technologies to communicate with vehicles, infrastructure, and other portable devices. I don't think that the connected car or vehicle-to-X communications is necessary for AVs to be successful initially, although some simple communications on things like traffic light status would be helpful.

### Vehicle-to-Vehicle

Vehicle-to-vehicle (V2V) communications use a wireless network where automobiles send messages to each other with information about what they're doing or things that are important about traffic conditions.

Instead of cars working independently, vehicles will be able to transmit vital information to nearby vehicles to improve the overall efficiency and safety of the roadways. V2V systems will use dedicated short-range communications (DSRC), which are two-way wireless channels that enable V2V-equipped cars to communicate with each other at roughly ten-times per second over short distances. On a busy highway, vehicles might send automated messages to each other communicating things like "Road is slippery," or "Ambulance coming!" or "Traveling 63 mph, road clear." It will give AVs situational awareness.

It can also be like the popular crowd-sourced application Waze, which warns drivers of accidents ahead, potholes, vehicles on the roadside, etc. It is unlikely that it will alert drivers of police ahead as Waze does. Some of the warnings communicated by other cars include forward collision warning where the cars ahead have an impending collision, emergency braking by the car ahead, lane change warning that a nearby car is changing lanes, intersection movement alerts that coordinates turns by all cars in an intersection, and a loss of control warning.

The U.S. Department of Transportation says car-to-car communication can help, as it has the potential "to significantly reduce many of the deadliest types of crashes through real-time advisories alerting drivers to imminent hazards." It could ease traffic congestion. Imagine if all the individual data points from thousands of cars went to a central hub. With so much real-time traffic data, transportation managers could adjust traffic light timing and redirect traffic to make rush hour flow more smoothly. Not only that, but each car could use the data to make individual adjustments.

It also presents privacy concerns. V2V data can reveal all kinds of things about you: where you're going and when, as well as your driving habits. Who has access to this data? And how might they use it? Liability could get complicated. V2V communications could be more vulnerable to hacking. The proliferation of

V2V communication, whether in human-piloted or driverless cars, gives malicious hackers a new opportunity.

### Vehicle-to-Infrastructure

Communications of AVs with traffic signals or other stationary devices is called vehicle to infrastructure (V2I). The application for this and the expected benefits are still unclear. There are different focuses too: is the information being transferred from the infrastructure to the vehicle, or is the vehicle communicating information to the infrastructure? There are also a few simple applications and much more complex applications.

I think the initial applications will be simple ones that communicate essential information to AVs. The most basic application is a traffic signal conveying its status to approaching vehicles: "Currently green for the next 60 seconds." This capability may lead to Signal Phase and Timing (SPaT). SPaT applications of V2I communications focus on the ability to coordinate driving speeds with traffic light patterns to be able to maximize fuel economy and speed. SPaT patterns can be used to provide the optimal traffic flow in high traffic areas.

Helpful information transmitted from the infrastructure to a vehicle includes: traffic signal information, stop sign and stop location information, lane closures, recommended speed, hazards/construction sites, intersection maps, and updates. These seem to be affordable and reasonable, and I expect metropolitan areas will implement these to be compatible with AVs, especially autonomous ride services.

The full extent of more complex V2I technologies' benefits and costs is unclear.

### Intelligent Stickers on Street Signs

One of the easiest and low-cost alternatives to having signs communicate with AVs is being explored by 3M, the maker of Post-It-Notes and Scotch Tape. They are testing stickers that appear to be transparent to the naked eye, but contain something like "barcodes" that autonomous vehicles can read.

In addition to defining the sign itself, these codes could contain GPS coordinates, traffic light warnings, distance to the curb, corner distances, etc. They are also testing this for work-zone signs and road-worker vests.

## 5G Communications

Simply put, 5G is the next "G" or "Generation" of wireless networks. Up until now 3G and 4G have allowed users to connect to the internet via their phones, with each new generation up to 10 times faster than its predecessor. But the next step, 5G, will not only benefit communication between people, but also between machines, promising significant benefits for the energy, healthcare, and automotive industries.

It will let more data move at higher speeds with lower latency and ultra-reliability, and it will be essential in supporting the billions of connected devices. The big difference with 5G is that autonomy requires split-second connectivity, which 5G can provide.

There are differing opinions on the importance of 5G for autonomous vehicles. Is 5G a requirement before AVs can be useful? Probably not. Dedicated short-range communications (DSRC), just discussed, can accommodate all necessary vehicle-to-vehicle and vehicle-to-infrastructure communications in modules that are already commercially available. To its credit, DSRC already boasts many desirable features for assisted and autonomous driving: it operates in fog or a snowstorm, can function at high speeds, and has a time delay of mere milliseconds. The wireless technology enables vehicles to send and receive brief digital packets of information about their whereabouts, intentions, and speed within a short to medium range.

And as I previously mentioned, AVs can operate successfully without the connected car. So, I don't expect AVs to <u>rely</u> on 5G communications.

Where 5G becomes very important is in mobile entertainment. Driverless cars will free everyone to be a passenger, and there will be an immediate demand for streaming video and related services that will require 5G communications. As mobile network operators begin to introduce 5G technology into their networks, they will focus on urban areas. 5G technology is expected to improve mobile wireless network capacity significantly and increase data speeds, allowing network providers to offer much more robust internet connections to devices.

## Electric vs. Hybrid Vehicles

Electric vehicles (EV), or battery-operated vehicles, are not the primary focus of this book, but this is an essential set of technologies for autonomous vehicles. There is a technology question on whether AVs will primarily be electric battery-operated or not.

Autonomous vehicles are technology-based platforms, and electrically powered cars are much more compatible than internal combustion engines (ICE) to being operated by software. Electric motors are more responsive to control and easier to integrate into an autonomous platform. In most cases, autonomous vehicles will be entirely new platforms built for the future, and since most experts expect electric to be the future drive system for cars, it makes sense to use electrically powered engines as the basis for the platform. Although some of the early AVs may modify existing vehicles.

Tesla set the stage for combining autonomous capabilities and electric vehicles. The technology companies expected to create autonomous vehicles are environmentally friendly; they don't want to build an ICE. Autonomous EV's also will encourage governments to enable favorable regulations for autonomous vehicles.

However, there several reasons why electric vehicle propulsion may <u>not</u> become the standard for early AVs.

The first is the significant power consumption of the computers and sensors in an AV. This power consumption could quickly drain the batteries in electric

AVs, limiting their driving range even further. An AV engineer in GM disclosed that the first-generation Cruise AVs consumed 3-4 kilowatts, which could significantly drain the 60kWh capacity of a Bolt. AV engineers are still wrestling with this issue, and power consumption is expected to improve. Power consumption is a definite concern for all the AV systems, starting with powering the computers in AVs.

The limited availability of charging stations may also be a limiting factor in the rapid growth of electric AVs. It will take some time for enough charging stations to replace gas stations before long-distance driving can be comfortable.

There may also be concerns about raw material constraints to making enough EVs in high volumes. Research by John Petersen stated that manufacturing 34.5 million electric vehicles using cobalt batteries would require about 276,000 tons per year (TPY) of cobalt. Optimistic cobalt growth forecasts only show 41,200 TPY available for EV batteries by 2025. Most companies in the EV industry are moving away from LMO (lithium manganese oxide) and LIP (lithium iron phosphate) batteries and embracing high-energy NMC (nickel manganese cobalt) batteries because of performance benefits. But LMO and LFP batteries use cheaper and more abundant raw materials: lithium, manganese, and iron. So, battery technology may be an essential issue in volume production.

As I stated previously, autonomous ride services (ARS) will be the first AV market, and it will grow rapidly. Ford is taking an interesting strategy for this market. It is developing an entirely new AV from the ground up, instead of retrofitting an existing model, and it is targeting this AV for initial use by ARS companies. This new AV targeted for release in 2021 will be a hybrid. Ford's reason for this is quite compelling: it wants to make this vehicle more profitable for ARS providers. A hybrid can drive much longer without refueling than EVs, and this can be very important to ARS providers where utilization is critical. Having a significant portion of its fleet down for an hour of recharging during peak periods could become a big problem. Ford's competitors (GM, Mercedes-Benz, and Nissan) are committed to a battery-operated AV.

Most likely electric AVSs will become the standard. Battery technology, AV systems power consumption, raw material constraints, and charging station limitations may be short-term issues. In first generation AVs, however, hybrid AVs may play a significant role.

# Chapter 7
# AV Markets and Company Strategies

Autonomous vehicles (AVs) will create several vast new markets with powerful companies vying for these opportunities. In this chapter, I attempt to describe the strategies that these companies may use to seize these opportunities. I want to note in advance: these strategic depictions are not based on any inside or proprietary information. They are based solely on publicly available information, combined with my experience in strategy for high-technology companies. Some of these companies have disclosed their AV strategy, although I expect most of these will evolve as the markets emerge. Others may not even know their specific strategy yet, but their realistic options and general direction seem clear. And a few haven't even declared their intentions as to what AV markets they may enter, or also if they will, but they are investing heavily in the technology, so their entry into the market is likely.

Before getting into individual company strategies, let's look at an overview of the new markets and industries created from the AV upheaval.

## New Markets and Industries

Significant new markets will be created by autonomous vehicles, and each of these new markets will create a set of new industries that serve these markets. As a semantical clarification, although the terms are sometimes interchangeable, I use the term *market* to refer to the total business opportunity for a collection of customers buying similar goods and services, and the term *industry* to refer to the companies that compete in that market.

### Autonomous Ride Services Market

As previously described, autonomous ride services (ARS) will be the first significant new market created as vehicles become *sufficiently autonomous*. This market will grow very quickly. I optimistically estimate the U.S. market to be $125-$150 billion by 2025 and $750 billion by 2030. Eventually, it may surpass

$1 trillion. The industries that support this market have several layers, and I expect that these layers will combine in different ways.

The *ride request and dispatch platform* sits at the top. This layer is mainly the ride-hailing app used by passengers to request a ride, and the underlying computer services to dispatch vehicles, bill customers, and manage logistics. The ride request and dispatch platform will be supported increasingly by artificial intelligence and analytics to predict rider travel demand levels and travel patterns. Ride request and dispatch may be the controlling point in the entire ARS market, as customers will initiate a ride based on their preferred service. The major ridesharing companies, Uber and Lyft in the United States, have established an advantage to attract customers to ARS, but there will be other factors to draw customers, such as the level of availability of autonomous vehicles in an area.

ARS will be deployed by individual metropolitan areas across the country. I refer to these as *municipal ARS fleet operations*. Each ARS company will launch its service in a targeted metropolitan area by placing a fleet of autonomous vehicles in that metropolitan area and providing rides within that metropolitan area. For example, Uber will initiate ARS in San Francisco with a fleet of 500 cars, and Las Vegas with 750 cars, and Sacramento with 600 cars. Lyft may match this with fleets in the same cities or other cities. Once a metropolitan area becomes "saturated" with enough ARS vehicles, competitors will be discouraged from investing in those locations. Given the potential profitability of ARS, I expect that this will be like a "land grab" or "gold rush" where the primary ARS competitors will invest to capture the most valuable metropolitan markets as fast as possible. Once most of the markets are served then there will be some consolidation and rationalization of the industry.

The next critical level of this industry structure is the *fleet investment* in autonomous vehicles; essentially this is who owns an AV fleet. This investment will be quite significant, probably in the hundreds of millions of dollars per municipal area. The ARS companies may own and control their fleets, probably Waymo and Apple will do this since they have sufficient capital. Lyft will most likely use one of its partners to finance the fleet, and it is unclear what Uber will do.

*Fleet management* is the final layer of this industry. Fleet management includes the maintenance, update, cleaning, and refueling of the fleets of autonomous vehicles in each metropolitan area. While municipal fleet operations will most likely manage this, I expect they will subcontract most of it to others. Fleet management will be an opportunity for car rental companies, such as AVIS and Hertz that have the experience and facilities to do it. It may also be an opportunity for car dealers and entrepreneurs. The big ARS companies won't have much interest in doing this themselves, and I don't expect this to be a very profitable industry.

As I previously discussed, the business model for ARS is expected to be very profitable. How that profit gets distributed across the layers of this new industry remains to be seen. However, if it is like most other industries, the layer that controls the market will be the most profitable. Most likely the top level – the ride request and dispatch platform companies who control the customers -- will be the controlling point. And this may come down to something straightforward: which ARS apps are most popular. If this is the case, then Uber and Lyft have the advantage of already having loyal customers using their apps. Apple has a unique advantage because it can automatically put its ARS app on hundreds of millions of iPhones. Google may have some similar success in deploying its Waymo ARS app across customers using the Android operating system, but Android users have proven to be very slow to update to new operating systems.

## Autonomous Vehicle Retail Market

Autonomous vehicles will also be purchased by individuals just as cars are today. I expect that the AV retail market will emerge very differently than the ARS market. It will follow much more of an evolutionary approach than the "land rush" of the ARS market. It will evolve in two ways. First, the level of autonomy in cars will increase over time. Several vehicles are already semi-autonomous, and these will increasingly offer more autonomous functions while still enabling the vehicles to be operated by a driver. Autonomous driving capabilities are typically offered as options or as particular models of cars, and I expect that this will continue. Eventually, these cars will be fully autonomous. Most car manufacturers will sell autonomous and traditional non-autonomous models of cars. Secondly, the market will evolve as people are willing to pay more for the autonomous functionality. This additional cost could be $10,000 to as much as $25,000 per vehicle initially, but will eventually come down.

The automobile retail industry in the United States has two distinct segments. *Auto manufacturers* design and manufacture the vehicles. *Auto dealers* sell and service the cars. In fact, this separation is currently required by law in most states. Tesla's strategy is to be a leader in this market by providing both. The traditional car manufacturers that make autonomous vehicles will probably continue to use their licensed dealers for sales and service.

There will be winners and losers in this industry. Companies that provide the best functioning autonomous cars will eventually win this market, but it will be a competitive battle for many years. Superimposed on this competition is not only the increasing market share for autonomous vehicles over traditional cars, but also, more importantly, the shrinking total market for new vehicles as ARS reduces individual car ownership, and with its higher vehicle utilization creates the need for fewer cars. Eventually, there will be fewer auto manufacturers.

The total retail auto market in the United States is almost $1 trillion today, but it will shrink by approximately 20% by 2025 and 40% by 2030. Semi-autonomous or autonomous cars will be an increasingly more significant percentage of cars sold. By 2025 most new cars will have some autonomous functions, and by 2030 most will be sufficiently autonomous.

*Automobile dealers* will go through a profound change with fewer cars sold and much more sophisticated maintenance challenges. I expect significant consolidation and an overall reduction in the number of automobile dealers.

## ARS Fleet Design and Manufacturing

The expected significant shift toward autonomous ride services (ARS) will create opportunities for ARS fleet design and manufacturing. Auto manufacturers currently rely on fleet sales, mostly to car rental companies, for 10%-35% of their total revenue. Fleet sales are considered a lower-profit market segment for them, and some manufacturers are starting to phase out fleet sales.

In the ARS fleet market, I'll break design and manufacturing into two different segments. I anticipate that there will be a new industry for *subcontract manufacturing* of AVs designed by others where the contracting company provides the overall design, the autonomous systems (sensors and computers), and software. This subcontract manufacturing approach is analogous to Apple designing iPhones and then using subcontractors like Foxconn to build them. If Apple enters the ARS market, as I expect, then subcontracting will most likely be its strategy for manufacturing AVs as well. Subcontract manufacturing of AVs will be part of the ARS fleet design and manufacturing market.

Some of the *traditional auto manufacturers* will prefer to do both the design and manufacturing, just as they do now. They would like to design the AVs, develop the autonomous systems platform including software, and get the value-added revenue and profit for doing that. For example, Ford would design and manufacture autonomous vehicles for Lyft. Lyft is a good example here because it is not investing a lot in developing autonomous systems, so it will rely on someone else to provide the complete AV.

Others cases are less clear. Will Waymo and Uber invest tens of billions designing and developing AVs, but then scrap all this and use someone else's AV platform? Not likely. Most likely they will subcontract to an automobile manufacturer to provide the "dumb" car platform to integrate with their intelligent AV systems.

Until there is some clarity on the eventual strategies, the make-up and size of the ARS fleet design and manufacturing market are impossible to predict.

## Technology and Component Supplier Markets

Companies that supply the technology and components for AVs will also create new markets. I classify these suppliers into three AV component markets: computing, sensors, and software.

The *computing market* includes companies that make the computer processing systems for AVs. This critical component may be referred to as microprocessors, supercomputers, processing platforms, etc. but they all provide basically the same function. They process (fuse) sensor data, position the vehicle and give commands to direct the vehicle along its route. Nvidia estimates that the market for AV computer processing will be $4-$5 billion by 2025.

In addition, there will be a *market for other components* that supplement the computing platform to provide specific functions. It's too early to estimate the size of this market since the AV architecture is still in process.

The *sensor market* consists of those companies that make the camera, radar, sonar, and lidar sensors for AVs. Each of these sensor types is a different segment of this broader market. In general, companies that make sensors compete in only one of the sensor market segments. Some estimate the AV sensor market to be as much as $15 billion in 2025 and $25 billion in 2030, but I think that may be unrealistically high as costs come down.

AV *software is a market*, but while there are some companies just focused on developing this software, companies that are developing complete AV systems do most of the software development. The software is the critical element of AVs, and it is expected to be one of the primary competitive differentiators.

## Supporting Markets

In addition to these primary markets, there is also a more general market for a wide variety of different industries that will support autonomous vehicles. This market includes HD maps, AV cleaning and maintenance, ARS dispatch systems, infotainment systems inside of AVs, test equipment for AV production, AV insurance companies, and many others. Some of these will provide exciting opportunities for new businesses. In general, the size and potential of these industries won't be apparent until the primary AV market is established.

# Technology Companies

Waymo (Google) is the pioneer of autonomous vehicles (AVs), and many experts recognize it as having a lead over everyone else in this market. It is one of two major technology companies in the United States that I expect will be significant competitors in the AV market. The other is Apple, which has not yet committed to this market but has given provocative indications of its interest.

The technology companies come into the autonomous vehicle market very differently than the car manufacturers. They don't have any manufacturing, automotive design, supply-chain, distribution, or service capabilities. While that is a significant constraint, it also frees them from the burden of continuing to invest in, manufacture, and support "legacy" vehicles. They can focus on autonomous vehicles. They also have significant experience and resources for developing sophisticated software, which is the key to autonomous driving.

## Waymo (Google/Alphabet)

Google is considered the pioneer in self-driving cars, starting its self-driving car project in 2009. Google's massive cash flows and scattered innovation "strategy" enabled it to invest heavily in what was a speculative initiative at the time. Many believe that Waymo (Google's AV  subsidiary) took advantage of this lead to develop the best AV technology. It certainly has more autonomous driving experience.

In late 2015, Alphabet hired John Krafcik, a former Hyundai North America executive, as the unit's first CEO, and in December 2016, it officially separated its self-driving research unit into its own company called Waymo. That was a clear intention by Google that it expected Waymo to become viable as a stand-alone business, rather than a mere research project. In fact, Morgan Stanley believes Waymo could be worth $70 billion on its own.

### Waymo Technology

Waymo developed its lidar system, a technology many believe is critical to the long-term success of self-driving cars. It claims to have lowered the cost of lidar sensors by 90%. Its AVs use three lidar sensors (near, medium, and long-distance detection), in combination with a 360-degree radar system and several proprietary camera-based sensors. This investment demonstrates that it's likely that Waymo will use its sensor hardware in addition to its software in the AVs it brings to market.

Its first-mover advantage gives Waymo the most self-driven miles of any company, with over 3 million miles driven on roads, along with over 1 billion miles in simulation by the end of 2017. This experience created continuing improvements. Its disengagement rate -- the rate at which a human must take over the vehicle -- dropped to only 0.2 disengagements per 1,000 miles in 2016.

### Commercialization

Waymo will focus on the ARS market opportunity. In the first stage of ARS, it looks like Waymo will retrofit vehicles currently produced by major car manufacturers. Waymo is using 600 Fiat Chrysler (FCA) Pacifica Hybrid minivans to pilot test its autonomous ride services. In June 2018, it announced that it would acquire up to 62,000 more, most likely over the next 2-3 years. Shortly before that announcement, Waymo also announced its intention to purchase up to 20,000 Jaguar I-Pace electric vehicles. If Waymo acquires these vehicles, it would have an ARS fleet of more than 80,000 vehicles. This is a significant portion of the 100,000 ARS vehicles that I estimate for the total U.S. ARS market by 2021.

It's unlikely that Waymo will sell AVs in the retail market since it would need to build significant sales and support infrastructure. However, there are some indications that it could license its AV technology to FCA as part of its partnership.

### ARS

Waymo's strategy is beginning to come into focus. It is entering the ARS market on its own. It has announced its intention to start providing ARS on a pilot basis without safety drivers by the end of 2018 in Arizona. It appears that Waymo will provide an ARS app and manage its own fleet of ARS vehicles.

It appears that Waymo is going to be very aggressive entering the ARS market. 80,000 ARS vehicles would translate into approximately $12 billion in revenue. Note, Waymo uses the same estimates as I do of 50 trips per-day per ARS vehicle. This could also translate into 50-80 different municipal markets.

Waymo will retrofit these vehicles by adding its own AV platform, including its sensor package, computing, and software. It has not indicated where or how it will do this retrofitting.

102

Waymo and Uber confronted each other in a bitter legal battle over Uber's alleged theft of Waymo's lidar technology, which was eventually decided in Waymo's favor. While this makes them unlikely partners, Uber's recent difficulties in testing its AVs may change this relationship. Uber has recently stated that it would be interested in working with Waymo on ARS. However, it is more likely that these two companies will be competitors instead of partners.

Waymo established a partnership with Lyft. The partnership's stated goal was to bring self-driving car tech to the mainstream through joint projects and product development efforts. This partnership aligns Waymo with Lyft in competition with Uber, which makes sense. However, Lyft also received funding from General Motors with the expectation it would partner with GM on self-driving initiatives, although this seems to be deteriorating too.

Waymo piloted its ARS business by offering the public rides in driverless Chrysler Pacifica minivans. Also, Google bought Waze for about $1 billion in 2013, and the service has since grown to 75 million users, up from the approximately 45 million users. Waze, which is not part of Waymo, is quietly transforming into a ride-sharing service that could eventually rival Uber and Lyft. It introduced a carpool service in several U.S. cities. Waze carpool encourages regular drivers to pick up other users along their route and is, in most cases, several dollars cheaper than Uber or Lyft.

In June 2017, Avis and Waymo announced an agreement with Avis to store and maintain Waymo's fleet of AVs in Phoenix. While this is a small isolated deal, Avis stock jumped 17%. If Waymo's strategy is in fact to launch its own autonomous ride service, it will need the infrastructure as Avis has for service.

Recently, Waymo's CEO stated that "We are a self-driving technology company. We've made it pretty clear we are not a car company".

### Waymo's (Google's) Strategy

Waymo has openly embraced the market opportunity for AVs, and it invested several billion dollars developing some excellent technology. Waymo wants to be a significant player in the AV market, and it won't be content to merely sell its self-driving software and sensors to help make other companies successful. Entering the AV retail market will be too formidable because it would need to create manufacturing, distribution, sales and service capabilities. This then leaves the autonomous ride services (ARS) market as its prime opportunity, which will be the largest initial market where it can use its technology to an advantage.

It looks like Waymo is going to enter the ARS market aggressively on its own and not partnering with Uber or Lyft. It intends to start providing this service initially in Arizona in late 2018 or early 2019. Then it intends to ramp up this service in more locations in Arizona, as well as in other locations over the next 2-3 years. Waymo's intended purchase of 80,000 cars to retrofit to be ARS vehicles shows that intends to ramp up its ARS business very rapidly.

Overall, Waymo has an early-leader advantage in AV technology. It will be the first to enter and create the ARS market, and it intends to rapidly expand its position in this market as long as there are no major failures or legal restrictions along the way.

## Apple

Reportedly Apple executives discussed building a car before the iPhone, but ultimately decided not to work on one.

Apple is always secretive about its product development strategy, but in June of 2017, Apple CEO Tim Cook spoke publicly about Apple's work on autonomous driving software, confirming the company's work. "We're focusing on autonomous systems. It's a core technology that we view as very important. We sort of see it (referring to autonomous vehicles) as the mother of all AI projects. It's probably one of the most difficult AI projects actually to work on." "There is a major disruption looming there," Cook later said on Bloomberg Television, citing self-driving technology, electric vehicles, and ride-hailing. "You've got kind of three vectors of change generally happening in the same time frame." In the interview, Cook was hesitant to disclose whether Apple will ultimately manufacture its car. "We'll see where it takes us," Cook said. "We're not saying from a product point of view what we will do."

### *Project Titan*

By 2014, Apple reportedly was working on "Project Titan," with upwards of 1,000 employees developing an electric AV at a secret location near its Cupertino headquarters. A few years later, Apple refocused the project under new leadership, laying off hundreds of employees who were working on the project. Apple transitioned to building an autonomous driving system rather than a full car.

In early 2017, Apple was granted a permit from the California DMV to test AVS on public roads using Lexus RX450h SUVs. In the spring of 2018, Apple increased the number of authorized test vehicles in California to 62, along with 83 permitted AV drivers. In June 2018, Apple was reported to have a deal with Volkswagen to customize its T6 Transporter vans to test a shuttle for its employees in late 2018 or early 2019.

Apple has been very aggressive hiring technical talent and executives from the auto industry, as well as Tesla, including battery experts. Several rumors about the Apple Car have included details suggesting Apple employees are working on the project at a top-secret location in the Bay Area. Apple leases several buildings at a Sunnyvale location, but it is also rumored to be operating a shell company at the site that received city permits for the construction of an "auto work area" and a "repair garage." The rumors of the car project in Sunnyvale may or may not be accurate, but based on past information, development on the car is indeed taking place at a secret location outside of the company's main campus. Apple is also rumored to be operating a secret vehicle research and development lab in Berlin with people from the German automotive industry, all with backgrounds in engineering, software, hardware, and sales.

Based on the significant increase in R&D and the rumor of 1,000 engineers,

Apple has most likely invested more than $1 billion in autonomous driving, and it is probably investing $300-$400 million a year.

### Partnerships

In May of 2016, Apple invested $1 billion in Chinese ride-hailing service Didi Chuxing. Didi Chuxing dominates the ridesharing industry in China with more than 80 percent market share. In July of 2016, Uber announced a merger between its operations in China and Didi Chuxing, further expanding Didi Chuxing's reach in the country.

Apple is also working on a processor explicitly devoted to AI-related tasks, known internally as the Apple Neural Engine, which would improve the way the company's devices handle tasks that would otherwise require human intelligence. An AI-enabled processor would help Apple integrate more advanced capabilities into devices, particularly AVs. A super-processing AI computer chip would provide another critical leg in its autonomous vehicle strategy. As Apple typical does with its platform strategy, this device would also provide an essential component for other artificial intelligence products.

Apple is also reportedly investing in automotive battery technology with Contemporary Amperex Technology Ltd. (CATL) based in China. The company is the third-largest battery manufacturer in the world, making batteries for electric buses, passenger vehicles, trucks, as well as energy storage.

### Apple's Strategy

I don't have any inside information on Apple's intentions in this market, but I think there is enough evidence and strategic logic to deduce its potential strategy. While apple hasn't confirmed its intention to create an autonomous vehicle, I believe that Apple realizes that this market opportunity is too big to ignore. Moreover, there have been sufficient hints in information in the public domain to support this.

Apple's autonomous vehicle strategy can be deduced by understanding what it does well and what it tends to avoid. Apple will bring to market a unique and exciting autonomous vehicle based on a proprietary platform it is currently developing. Apple will design most of the vehicle especially a passenger entertainment focused interior. It will create a proprietary platform of sensors, computing, and software. Apple might provide its own graphics processing computer, but it may use one from Nvidia to start. It will eventually subcontract vehicle manufacturing, just as it does today with its iPhones and iPads.

I don't expect Apple to enter the AV retail market or sell directly to end customers. It is just too difficult to build a sales, distribution, and service organization. Instead, it will enter the ARS market aggressively. Apple has a significant advantage in merely automatically downloading an Apple ARS app to hundreds of millions of iPhones and letting customers use their existing iTunes account for charges. It's unclear how much of the ARS operations Apple will finance and outsource.

Like Waymo, initially Apple will most likely customize cars currently produced, turning them into autonomous ARS vehicles. Then I expect that it will custom design them in the second stage of the ARS market.

I also expect that Apple will leverage its video and music resources to provide entertainment in its ARS vehicles. Passengers will automatically have access to all their purchased and streaming Apple content.

Assuming Apple does enter the AV market, particularly autonomous ride services, I expect that Apple will capture significant market share, with incremental revenue in the range of $50 billion to $100 billion per year by 2030.

### Microsoft

Microsoft is not a major player in autonomous vehicle development. It has some research into self-driving cars, but its initial strategy appears to focus on collaborations, such as a deal with Volvo that will see the companies collaborating in autonomous vehicle R&D and leveraging Microsoft's HoloLens technology. In March 2016, Microsoft and Toyota also  announced the expansion of their five-year-old partnership to support Toyota's research in robotics, AI, and self-driving car development. Microsoft's strategy seems focused on providing automakers with technological assistance along with growing its Azure cloud business, as opposed to developing an AV.

## Technology-Based Car Manufacturer

Tesla is in its own category, which I call technology-based car manufacturer, as there is nobody else like it in the autonomous vehicle industry. It has a different strategy and is using a unique approach.

### Tesla

Tesla first gained widespread attention following its production of the Tesla Roadster, the first electric sports car, in 2008. The company's second vehicle, the Model S, an electric luxury sedan, debuted in 2012 and is built at the Tesla Factory in  California. The Model S was then followed in September 2015 by the Model X, a crossover SUV. In March 2016, Tesla introduced the Model 3 for a base price of around $35,000, with shipments delayed until 2018.

Tesla's product strategy emulates other successful technology companies. It created a unique technology platform with its battery, power train, and autonomous driving technology. It initially entered the market with an expensive, high-end product targeted at affluent buyers. Then as the cost of the platform was reduced, it moved into larger, more competitive markets at lower price points. Elon Musk stated, "New technology in any field takes a few versions to optimize before reaching the mass market, and in this case, it is competing with 150 years and trillions of dollars spent on gasoline cars."

Tesla invests almost $1 billion in R&D annually and has invested more than $4 billion since inception. It's difficult to determine how much of this is invested in autonomous driving since it also invests in electric car and battery technologies. In the end, it's all invested in an electronic autonomous vehicle. Tesla Inc. is working with Advanced Micro Devices to develop its own artificial intelligence processor that, if successful, could replace the Nvidia processor it currently uses.

### Tesla Infrastructure

Because Tesla doesn't have auto dealers to sell its cars, it created a unique sales, service, and distribution channel. It owns and operates more than 200 stores and galleries in the United States, which sell directly to customers using the internet to place customer orders because 48 states have laws that restrict or ban manufacturers from selling vehicles directly to consumers. Outside of the U.S. Tesla sells cars directly to consumers. Dealership associations in multiple states have filed numerous lawsuits against Tesla to prevent the company from selling cars directly. However, the Federal Trade Commission recommends allowing direct manufacturer sales, so those restrictions eventually may be lifted.

The entrenched auto dealers will keep battling Tesla's direct sales model for a long time and in many ways. For example, the Texas Commission on Environmental Quality initiated a $2,500 electric vehicle rebate program, but won't provide it for Tesla cars: "The vehicle must be purchased or leased from a licensed new vehicle dealer or leasing company in Texas. Vehicles purchased directly from the manufacturer or an out-of-state dealer not licensed to sell or lease new vehicles in Texas are not eligible for a rebate."

Tesla is struggling to keep up with service centers as it expands its customer base. Because of the proprietary nature of its cars, all major service must be done by Tesla. In 2017 it had approximately 150 service centers, and many customers complained of long wait times. Tesla previously used a valet service to pick up cars but now sends technicians to the owner's home to do much of the maintenance.

To make its electric cars feasible for longer-range driving, Tesla created a network of high-powered Superchargers located across North America, Europe and Asia. It also operates a Destination Charging program, under which shops, restaurants and other venues provide fast charging stations for their customers.

Perhaps Tesla's biggest challenge is manufacturing cars in sufficient volume. While it had 300,000 deposits on reservations for the Model 3 sedan, by the end of 2017 it was only able to ship 1,550 cars, and it continued to struggle with production well into 2018. Tesla is much more vertically integrated (including battery manufacturing) than other car manufacturers, which gives it more control over manufacturing, but also makes it more challenging to ramp up manufacturing.

### Tesla Autopilot (Autonomous Vehicle Technology)

Right from the beginning, Tesla's strategy was to build autonomous cars, and it *may* be making rapid progress, although some question the reliability of its claims. Tesla Autopilot 1.0 provides semi-autonomous driver assist in Tesla vehicles. These vehicles are equipped with a camera mounted at the top of the windshield, a forward-looking radar in the lower grill and ultrasonic acoustic location sensors in the front and rear bumpers that provide a 360-degree buffer zone around the car. This equipment allows vehicles to detect road signs, lane markings, obstacles and other vehicles. In addition to adaptive cruise control and lane departure warning, a "Tech Package" option allows this system to enable semi-autonomous drive (called Summon) and parking capabilities (called AutoPark). These features were activated via over-the-air software updates. The AutoPilot system as of version 8 uses the radar as the primary sensor instead of the camera. In the summer of 2018, Tesla is expected to release major upgrade of Autopilot.

### Autopilot 1

Autopilot was first offered in 2014 for Tesla Model S, followed by the Model X. Autopilot was part of a $2,500 "Tech Package" option. At that time Autopilot features included semi-autonomous drive and parking capabilities. Initial versions of Autopilot were developed in partnership with the Israeli company Mobileye, but Tesla and Mobileye ended their partnership in July 2016 after some bitter disagreements.

In 2015, Tesla released software package version 7.0 with Autopilot to its customers, but later removed some self-driving features to discourage customers from engaging in risky behavior. In 2016, Elon Musk announced Autopilot Firmware 8.0, that processes radar signals to create a coarse point cloud like lidar to help navigate in low visibility conditions, and even to 'see' in front of the car ahead. Autopilot, as of version 8, uses radar as the primary sensor instead of the camera. Autopilot 8.0 was later updated to have a more noticeable signal that it is engaged and requires drivers to touch the steering wheel more frequently. By November 2016, Autopilot had operated actively on hardware version 1 vehicles for 300 million miles (500 million km) and 1.3 billion miles (2 billion km) in shadow mode. Autopilot 1.0 is no longer sold in new cars, but continues to be updated. It was updated with autopilot 2.

### Autopilot 2

Tesla was supposed to release autopilot 2 by the end of 2017, but it was delayed.

### Hardware 1

Vehicles manufactured after late September 2014 are equipped with a camera mounted at the top of the windshield, forward looking radar (supplied by Bosch) in the lower grill, and ultrasonic acoustic location sensors in the front and rear bumpers that provide a 360-degree view around the car. The computer is the Mobileye EyeQ3. This equipment allows Model S to detect road signs, lane markings, obstacles, and other vehicles. Upgrading from Hardware 1 to Hardware 2 is not offered as it would require substantial work and cost.

### Hardware 2

As of October 2016, all Tesla vehicles came with the necessary sensing and computing hardware, known has hardware version 2 (HW2), for future fully autonomous operation, with software being made available as it matures. Hardware 2 includes an Nvidia Drive PX 2 GPU. Tesla claimed that Hardware 2 was 40 times faster than Hardware 1 to allow full self-driving capability. The sensors in HW2 include 8 surround cameras and 12 ultrasonic sensors, in addition to forward-facing radar with enhanced processing capabilities. The radar was claimed to be able to see beneath the vehicle in front of the Tesla to observe the vehicle in front of that vehicle.

The first release of Autopilot for HW2 cars included adaptive cruise control, auto-steer enabled on divided highways, and auto-steer on 'local roads' up to a speed of 35 mph. Firmware version 8.1 for HW2 began in June 2017 that included many new features including a new Autopilot driving-assist algorithm, full-speed braking and handling parallel and perpendicular parking. A new release is planned for the summer of 2018.

### Partnership with Mobileye

In September 2016, Tesla and Mobileye dissolved their partnership and supply contracts and left Tesla to develop its own driving-assist Autopilot technology as a complete system. Tesla said at the time that Mobileye, "attempted to force Tesla to discontinue this development, pay them more, and use their products in future hardware," while Mobileye disputed the claim. Instead, Mobileye made a claim that Tesla was, "pushing the envelope in terms of safety," and wanted Tesla to lower the abilities of Autopilot until it was proven to be safer.

Musk claims that engineers at Tesla recreated the technology powering the Mobileye chip in just six months, which was a fraction of the time that Mobileye had spent millions to create it. Mobileye was acquired by Intel for $15.3 billion.

Tesla claims that all cars built after October 2016 have the necessary hardware to allow full self-driving capability at a safe level (SAE Level 5). The hardware includes eight surround cameras and twelve ultrasonic sensors, in addition to the forward-facing radar. Also, the system will operate in "shadow mode" (processing without acting) and send data back to Tesla to improve its abilities until the software is ready for deployment via over-the-air upgrades.

### Timing for Autonomous Driving

In 2017, Elon Musk clarified plans for autonomous driving and now predicts that true level-5 autonomy is about 2 years away, making it 2019. While Musk said that the "full self-driving capability" option on the second-generation Autopilot will eventually enable level-5 autonomous driving, which means fully autonomous in all conditions, Tesla also specified that it is dependent on software validation and regulatory approval.

However, there are reports of disagreement within Tesla about its software development. There has been much turnover in the leadership of Tesla's AP 2 development, suggesting that its release target dates may be overly optimistic. Some believe that Musk has promised something that Tesla can't deliver by the dates promised, and possibly not at all.

### Reliance on Vision Technology

Tesla's AVs rely on vision systems, while most other companies utilize lidar and radar systems as well. Musk has reiterated Tesla's vision-based approach – reassuring current owners that autonomy can be achieved using just cameras: "Once you solve cameras for vision, autonomy is solved; if you don't solve vision, it's not solved ... You can absolutely be superhuman with just cameras."

This is a lower-cost strategy that makes a reasonably priced AV possible, but it requires much more processing power, and Tesla may find that it will need to upgrade the processing power in the car for this to work. In addition, there have been reports in a tear-down analysis of Tesla's processor that it may only be a partial implementation of an early version of the Nvidia Drive PX2.

So, in danger of oversimplification, one could argue that the race towards AVs is between getting the cost of lidar down versus improving the algorithms of the computer vision driving on non-lidar-based systems like Tesla's.

### Tesla Network

Tesla also plans to launch a ridesharing service, Tesla Network, for Tesla's self-driving cars. It places restrictions on its customers restricting them to using the Tesla Network for ridesharing. Most likely this will be implemented with a Tesla ridesharing app using autonomous cars provided by its customers. The theory is that customers will want to have their Tesla AVs go off on their own and do some work to earn money for the owners when they aren't using their cars.

Tesla doesn't have the capital required to invest in its own ARS business like Apple, Uber and Google, and possibly some of the auto manufacturers. While this may be a novelty segment of the ARS market, I don't see it becoming a major segment. ARS will require a reliable number of available cars, and many Tesla owners won't want to deal with dirty or damaged cars.

### Tesla Fatal Accidents

Tesla vehicles have been involved in two fatal accidents. While everyone should expect that there will continue to be some fatal accidents with AVs, any early accidents need to be analyzed. On March 23rd 2018, a Tesla Model X crashed into a freeway divider killing the driver. According to the NTSB the vehicle accelerated from 62 to 71 MPH prior to the crash with no braking or evasive steering. Autopilot was engaged and the speed was set for 75 and the driver's hands were not on the steering wheel for the prior 6 seconds. In reviewing the video of the location and the video of a subsequent Tesla driver who tried the same location, it is possible that the vehicle misread the painted lines that veered off to the left and moved the vehicle in that direction, accelerating as the car in front was no longer in its path. Tesla pointed out that the crash barrier was defective because it had been impacted by a prior vehicle, and that overall AVs will prevent more fatal accidents.

As previously mentioned, a Tesla was also involved in an earlier fatal accident when it accelerated into a truck because it couldn't distinguish the truck form the shy.

### Tesla's Strategy

Tesla was one of the early pioneers in AVs, and it may go on to be very successful. Certainly, the high valuation of its stock price indicates that many investors believe that it will. Tesla's stated objective is to be the leader in selling autonomous (Level 4-5) cars to the end customer. Its strategy is to progressively build autonomous capabilities into the software for cars that it is selling primarily as electronic vehicles. With continuous software releases it can upgrade cars that it previously sold to Level 4 and then to Level 5, if it gets the sensor and computing requirements correct in the first place – and there is significant risk that it won't.

Tesla can follow this strategy of selling electronic vehicles and then directly upgrading them with autonomous software capabilities. It could be a sound strategy for early low-volume penetration for the emerging market for autonomous vehicles. Given manufacturing constraints, I expect that Tesla will have sold approximately 500,000 cars by the end of 2020 that are capable of autonomous driving. Since it sold most of these cars as electronic vehicles rather than autonomous vehicles, not all of the customers will turn on or use the autonomous software capabilities. If half turn their cars into autonomous vehicles, and if its autonomous technology works, Tesla could have 250,000 autonomous vehicles on the road by 2021, and from then on it will rely mostly on customers who buy its cars for autonomous driving, probably selling 200,000 – 300,000 per year, since the customer-owned (non-ride services) market will evolve more slowly.

I am a big fan of Tesla, but it faces some major strategic risks. Tesla may have made a strategic mistake with a conflicting core strategic vision. Is its objective to build electric vehicles for the masses or to be a pioneer in autonomous driving? These may be inconsistent objectives. The introduction of the 3 series creates a more-affordable electric vehicle, *Tesla may have conflicting core strategic visions between an EV* but the evolution of autonomous driving *for the masses and an AV.* initially will require more expensive vehicles, because the cost of the autonomous technologies is expensive. This dual strategy has forced Tesla to adopt a lower-cost strategy for its autonomous driving capabilities, particularly not using lidar, which may create a major technical risk.

Tesla needs to get the autonomous-driving hardware correct when representing it as an electric vehicle eventually capable of being autonomous. If the sensors or computing power isn't sufficient, there could be a big backlash. It needs to overcome the sales and service network challenges. Scaling the service organization will be a challenge. Tesla needs to build quality cars in volume. Finally, Tesla will require an enormous amount of capital, which it doesn't have. Acquiring SolarCity adds to its capital problem, as well as being a distraction.

I don't see Tesla Network as being successful in the autonomous ride services market. It relies on geographically distributed Tesla owners providing their personal cars, when the key to success in this market is a fleet of AVs in each geographical market.

Overall, my assessment is that Tesla's strategy may enable it to be an early pioneer in autonomous driving, capturing an early share of the retail AV market. However, most of the early growth and success in autonomous vehicles will be in autonomous ride services market. The traditional car manufacturers are also catching up fast with Tesla, and once they have viable AVs, they will be capable of manufacturing them in volume and provide proven sales and service capabilities. Tesla faces some significant challenges, but given Elon Musk's strong leadership, I wouldn't bet against it.

## Ridesharing Companies

There are two primary ridesharing companies in the United States and a host of smaller ones because there are few barriers to entry. It is well known that the current ridesharing model of using contract drivers with their own cars is a temporary step on the way to something better. As it is today, ridesharing is a low cost to the ridesharing companies, but it is an unprofitable business model.

SharesPost research estimates that ridesharing companies have raised more than $30 billion in private capital since 2010. Today, the top-5 global ridesharing companies – Uber, Didi-Chuxing, Lyft, Ola, and Grab – have a combined market capitalization of roughly $120 billion (based on most recent investment valuations). Ridesharing companies have a large and expanding market opportunity, and benefit from significant secular and demographic advantages.

*All ridesharing companies want to leverage their customer base into ARS.*

The ridesharing companies all have the same long-term objective: to create a presence in ridesharing using contract drivers and to transfer that customer base and market penetration into autonomous ride services, which are much more lucrative. They see this as the Netflix model. Netflix entered its market by providing a service of mailing DVDs to customers as a subscription service. Then it created the video streaming market and transferred that customer base to that more profitable technology.

There are a couple problems in applying the Netflix strategy to ride-sharing. First, Netflix never started out with that strategy. It saw the opportunity after it had grown the business. Netflix was perceptive enough to see video streaming as a threat and opportunity and aggressively developed video streaming technology. Second, to keep with the same analogy, the ridesharing companies could be following the Blockbuster strategy. Blockbuster was too encumbered with its brick and mortar stores to move into the video streaming. Will the ridesharing companies be too encumbered with their driver networks?

Let's look at each of the two primary American ridesharing companies and their strategies, starting with Uber.

# Uber

Uber's original strategy was to provide a form of limo service to be "everyone's private driver." Now it tries to be "transportation as reliable as running water, everywhere for everyone." What started as an app to request premium black cars in a few metropolitan areas has completely changed the  way society uses transportation. Today, consumers can press a button on their mobile phone to get a ride, or get food delivered, or deliver a package – no matter what they want, when they want it, or where they want it.

Uber launched its service in San Francisco in June 2010 as "Uber Cab", but within 90 days, the company received a "cease and desist" letter from the San Francisco Municipal Transportation Agency, and dropped the word "cab" from the company name.

One of Uber's defining features is its ease-of-use. Riders simply press a button in the mobile app, and Uber matches the rider to the closest driver. The company believes that the network effects of its business model hinge upon Uber being the most reliable and convenient service. Further, the network effects between drivers, passengers, and information shared across trips, creates a direct feedback loop into improving the value in the marketplace.

Over the past three or so years, Uber has raised roughly $12 billion in funds and expanded its geographic footprint more than fifteen-fold. The opportunity provided by ARS is the major objective for its investors. Uber has grown to approximately $40 billion per year in billings, and it continues to grow rapidly.

### Autonomous Driving

Uber's founder, Travis Kalanick, said he views the race to bring self-driving technology to market as "existential" to Uber's future. The company has poured hundreds of millions of dollars into catching up with Google and others. Uber opened its Advanced Technologies Center in Pittsburgh, its center for self-driving-car research, in 2015.

In May 2016, Uber revealed its in-house autonomous prototypes for the first time, showing off retrofitted Ford Fusions from its Advanced Technologies Center (now Advanced Technologies Group). Uber also acquired Otto, a self-driving trucking start-up, for more than $680 million.

Uber has an agreement with Volvo to purchase 24,000 SUVs between 2019 and 2021 that it will convert to AVs. The estimated cost of the purchase is approximately $1 billion, although the total cost could be much higher when autonomous driving hardware is included. This agreement indicates several elements of Uber's potential strategy:

1. It appears to be making the investment in ARS vehicles, instead of relying on another company to make the investment and own the vehicles.
2. It suggests an initial roll-out time frame of 2019-2021, most likely loaded more at the backend.

3. It has chosen an SUV as the model for its first autonomous ride services vehicle.

### *Uber Fatal Accident*

The tragic death of a pedestrian hit by an Uber autonomous vehicles (AV) on March 18th provides some important lessons. Elaine Herzberg, 49, was walking her bicycle far from the crosswalk on a four-lane road in the Phoenix suburb of Tempe about 10 PM when she was struck by an Uber autonomous vehicle traveling at about 38 miles per hour. The Uber Volvo XC90 SUV test vehicle was in autonomous mode with an operator behind the wheel.

"The pedestrian was outside of the crosswalk. As soon as she walked into the lane of traffic she was struck," Tempe Police Sergeant Ronald Elcock told reporters at a news conference. The video footage shows the Uber vehicle was traveling in the rightmost of two lanes. Herzberg was moving from left to right, crossing from the center median to the curb on the far right.

In viewing the video of the accident, it was clear that a human could not have reacted in the 2 seconds from when the victim was visible. However, it was subsequently determined that the Uber AV's lidar detected her a few seconds earlier when she was in the dark and not visible. The software on the AV didn't identify her as a human with a bike immediately and thought it could just be a harmless object. Some AV experts see this as a failure of the Uber software. Uber suspended its AV testing for a period of time following the accident.

### *Uber Strategy*

Uber's basic strategy is clear: it intends to build a very large, but unprofitable, ridesharing business, and then convert its loyal customer base to very profitable ARS. It appears that Uber intends to develop and use its own proprietary AV technology and own its own fleet of AVs, although the recent accident may have called this into question.

Overall, I expect Uber to be a major competitor in ARS due to its dominant market ridesharing market share in the U.S. It has a leadership position in ridesharing and a recognizable brand for its app. There will also be more competition in the ARS business than there is in ridesharing.

## Lyft

Lyft is a ridesharing company based in San Francisco, California. It develops, markets and operates the Lyft car transportation mobile app. Launched in June 2012, Lyft operates in approximately 300 U.S. cities, and provides 18.7  million rides a month. The company was valued at $7.5 billion as of April 2017 and has raised a total of $2.61 billion in funding.

Lyft was launched in the summer of 2012 by Logan Green and John Zimmer as a service of Zimride, a long-distance ridesharing company the two founded in 2007. Zimride focused on ridesharing for longer trips, often between cities, and linked drivers and passengers through the Facebook Connect application.

Whereas Zimride was focused on college campuses, Lyft launched as an on-demand ridesharing network for shorter trips within cities.

*Partnerships*

Lyft has established several partnerships with companies that want to provide AVs.

In 2016, GM made a $500 million investment in Lyft. At the time, there were rumors that GM made an offer to acquire Lyft, but was turned down. GM said at the time that it would work with Lyft to develop autonomous ride services. However, these intentions didn't materialize because of differences. GM president Dan Ammann joined the board of Lyft shortly after the investment was made, but he subsequently resigned from the board. In June 2018, GM stated that it had no active projects underway with Lyft. It currently appears that GM will develop its own autonomous ride sharing service and compete with Lyft.

Ford is also working with Lyft to deploy its self-driving cars on the Lyft's service platform by 2021. They also announced intentions to work together developing autonomous technology. Ford is joining Lyft's open platform. Ford will use Lyft's ride-hailing service to test and develop software and will eventually deploy AVs.

In May 2017, Lyft and Waymo launched a self-driving vehicle partnership to compete with Uber. Lyft, said this deal would accelerate its vision for transportation and Waymo said the partnership would let its technology reach "more people, in more places".

In June 2017, Lyft announced a partnership with Boston-based startup Nutonomy (subsequently acquired by Aptiv) to eventually put "thousands" of on-demand, autonomous vehicles on the road. They intend to launch a limited pilot in Boston, in which Lyft users will be able to hail one of Nutonomy's driverless vehicles by using Lyft's app.

Also in June 2017, InMotion, a venture capital fund backed by Jaguar Land Rover, announced a $25 million investment in Lyft. As part of the deal, Jaguar Land Rover will supply Lyft with a fleet of Jaguar and Land Rover vehicles, while the U.K.-based automotive company develops and tests self-driving cars.

Lyft stated that these partners were non-exclusive and that each is unique. A Lyft spokesperson said, "We are not disclosing the details of the work we are doing with each partner. Overall, we're partnering with leaders in this space who share a vision of solving transportation issues and positively impact the future of our cities."

Lyft's rationale for these partnerships became clearer when it officially announced that it will develop a suite of hardware and software that will allow any manufacturer to turn its vehicles into autonomous vehicles. Lyft intends to enable its partners—including GM, Waymo, nuTonomy, and Jaguar Land Rover—to outfit their cars with its autonomous technology. The toolkit will offer mapping software, physical interfaces for drivers and passengers, path planning, and other necessary components of autonomous driving.

### *Autonomous Vehicle*

Lyft is making some investment in autonomous driving, but it is much more limited than others. Lyft has a self-driving division called Level 5. Based in Palo Alto, Level 5 will house hundreds of engineers that will work on developing an open self-driving system. Although it's unclear how Lyft will work with its various partners to implement self-driving technology into its network, these collaborations will help better position the ride-hailing company to take on the Uber.

### *Lyft Open Platform*

Lyft has created a first-of-its-kind open platform for AV providers. The platform will be used by AV manufacturers to provide ARS through Lyft.

It will give them access to potentially millions rides each week. The platform will provide vehicle dispatching that deploys when there's only the highest confidence in routes and conditions. In addition, it will provide access to an exclusive set of data based on real-life scenarios to inform development. Partners will be able to deploy their AVs quickly using Lyft's API.

### *Lyft Evolution to AVs*

In "The Third Transportation Revolution: Lyft's Vision for the Next Ten Years and Beyond." Lyft describes its strategy to roll out self-driving cars in three phases, the first of which will likely be available by 2018. In that phase, semi-autonomous cars will be made available to Lyft users, but will only drive along fixed routes that the technology is

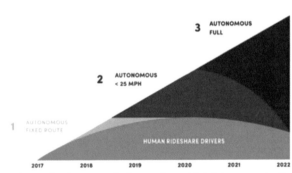

guaranteed to be able to navigate. In the second phase, the self-driving cars in the fleet will navigate more than just the fixed routes, but will only drive up to 25 miles per hour. As the technology matures and the software encounters more complex environments, Zimmer wrote, cars will get faster.

The third phase, which according to the chart the company expects to be sometime in 2021 or 2022, will be when more Lyft rides will be completed by a fully autonomous car.

### *Lyft Strategy*

Lyft is making some investment in developing autonomous vehicles, but not as much as Uber. Its strategy appears to rely on the technology development of others. Lyft is taking a collaborative approach while Uber is taking a proprietary one. Uber employs teams of engineers to work on developing its own self-driving technology, Lyft is focused on building a network of partnerships instead.

Unlike Uber, which looks like it will design and manage its own autonomous fleet, Lyft has a strategy of providing its autonomous ride services platform as an open platform for use by AVs developed by others. It's unclear who will finance, own, and manage the fleet of vehicles, but Lyft's general strategic direction seems strong as long as its partners let it control the customer. It will control the entry point and brand for its riders and may be able to extract high profit margins. It would also continue to have the future flexibility to become more vertically integrated if it wants, and it would also be a prime acquisition candidate.

## Automotive Manufacturers

Almost every automotive manufacturer is investing in autonomous vehicles, although some much more than others. Those investing more are trying to gain early mover advantages, while the others will try to be fast followers.

As a group, the current car manufacturers will lose, because the emergence of the autonomous ride services market will greatly reduce the number of cars manufactured worldwide. However, within this group several winners will emerge to become major players, if not leaders, in manufacturing and selling autonomous vehicles. In addition, it is likely that a few of the current automotive manufacturers may become subcontractors to build a high volume of cars designed by Apple, Waymo, Uber, or Lyft for autonomous ride services.

Auto manufacturers will general follow a somewhat similar strategy to varying degrees of success. They will progressively build autonomous capabilities into their cars, progressing from Level 2 to Level 3 to Level 4 and finally to Level 5 over time. In fact, many cars are already at Level 2 with partial automation. They will all follow this general strategy by necessity since they are already building, selling and servicing cars and have a loyal customer base to progressively upgrade to autonomous driving. It's impossible at this early stage to identify the winners and losers among current automotive manufacturers since their strategies are not yet clear (even to themselves in many cases).

There are too many auto manufacturers to try to understand each of their individual AV strategies, so I'll concentrate on the following.

### General Motors (GM)

GM believes that it is a leader in the development and commercialization of AVs. In 2017, GM purchased the AV startup, Cruise Automation, for $1 billion, making it the central part of its autonomous vehicle strategy. Following that acquisition, it also made several other key technology acquisitions.

GM is pursuing a broad solutions strategy for AVs, as is shown in one of its charts below describing its AV strategy, in which claims that it is the only company simultaneously developing an integrated strategy. This integrated strategy includes AV software development, HD mapping, proprietary sensors, and redundant hardware systems, combined with the processes to build AVs in mass production, and provide customer support.

| GM is the *only* company parallel iterating a *totally integrated* solution | | | | |
|---|---|---|---|---|
| Self-driving software "brain" | Deep simulation capability | HD Mapping and Routing | Proprietary AV sensors | AV-specific redundant hardware systems |
| Core EV platform | Automotive safety and durability validation | Cyber-security and electrical architecture | Vehicle connectivity and data collection | AV-specific vehicle design |
| Operations infrastructure | Large scale production readiness | UX interfaces (in car & app) | Customer support & remote assistance (OnStar) | Total cost optimization |

### Semi-Autonomous Capabilities

GM introduced semi-autonomous features in the 2018 Cadillac CT6. This is the company's first car with Super Cruise, a technology that can take over during highway drives. Cadillac maintains that this is the first production car with "true" hands-free driving. Instead of requiring that drivers keep their hands on the wheel, Cadillac uses an infrared camera only that tracks the drivers head to make sure they are paying attention.

Super Cruise uses an extra level of mapping accuracy recorded by lidar (but the CT6 doesn't have lidar itself) in addition to the usual onboard sensors and GPS. Cadillac only allows these features to operate on divided, limited-access highways with clear entrance and exit ramps. The CT6 is expensive, costing more than $75,000.

### GM AV Vehicle

GMs initial AV product is being developed using the Chevrolet Bolt EV. The autonomous version of the Chevrolet Bolt is the second generation, capable of handling nearly all road situations on its own without driver intervention. It is equipped with the latest array of equipment, including cameras, radar, sensors and

other hardware designed and built by GM and its suppliers. The automaker had already built about 50 Bolt AVs that were retrofitted with the specialized sensors needed to drive autonomously, giving it a total of 180 vehicles it can test and refine.

GM believes this is the first mass-produced, autonomous high-volume car. It has "full redundancy" throughout the autonomous system, so that it's ready mechanically and from a sensor and software perspective to "fail operationally and be safe." The vehicle is based on a third-generation Cruise self-driving platform. The new AV Chevrolet Bolt is being produced at the automaker's Orion, Michigan facility. This is a purpose-built vehicle, which is better than retrofit vehicles that are more difficult and more expensive to build and continually need to be fixed. While it looks very much like the current Bolt EV, 40% of the parts are new, and most of those are focused on redundancy of parts and systems.

In January 2018, GM showcased its fourth-generation AV. It has no driver controls and no steering wheel. This is clearly a commitment to fully autonomous vehicles that has not yet been seen in many others. GM Cruise has been able to reveal four iterations of its AV in a short period of time, reflecting its significant commitment and investment in AVs.

When in operation, GM says the vehicles will travel in geo-fenced areas, summoned by smartphones. Customers will use a mobile app to request a ride, just like they use ridesharing today. They will be able to control the experience — their customized climate control and radio station settings will be sent to the vehicle ahead of when they access their ride. GM expects that its AV will be available sometime in 2019, and it has stated repeatedly that these vehicles will be able to roll off the line in quantities of hundreds of thousands per year.

### GM AV Investment

GM is making considerable investments in AV development. The acquisition of Cruise Automation for $1 billion was its largest. GM also acquired lidar technology company Strobe, Inc. and Strobe's engineering talent joined GM's Cruise Automation team to define and develop next-generation lidar solutions for AVs. GM expects Strobe's lidar technology will significantly improve the cost and capabilities of its vehicles so that it can more quickly accomplish its mission to deploy AVs at scale. GM also invested in Nauto, following other automakers such as Toyota and BMW. Nauto uses camera systems to watch both the road ahead and the driver.

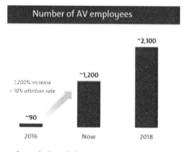

With these acquisitions, combined with the other core investments in AV development, GM is committed to a very large investment in AV development. GM expects to have approximately 2,100 employees involved with AV development in 2018. Based on this employee number, GM is probably investing at least $350 million, and probably more like $400 million annually in AVs.

### Autonomous Rides Services

General Motors is targeting autonomous ride services (ARS) as an initial market for AVs.

GM acquired car-sharing service Maven. General Motors subsequently launched its car-sharing service Maven in Ann Arbor, Michigan, though GM plans to move into other cities. There is a bit difference between Maven, which is a car-sharing service like ZipCar, and the ridesharing services provided by Uber and Lyft.

Cruise Automation, a tech startup acquired by GM last year, has also begun testing an app-based service for its roughly 150 employees to be shuttled around Silicon Valley in self-driving cars.

It appears that GM knows that ARS will be the first market and primary battleground for AVs, it has several strategies to get there, but none of them are sure things. GM admits this saying "To get us to the future faster, we're prepared to go with one partner, many partners, or no partners at all".

In its boldest AV move and an indication of a major ARS strategy, GM effectively spun-off Cruise as an AV ridesharing business with the investment of $2.25 billion by SoftBank. GM will also invest an additional $1.1 billion with SoftBank owning almost 20% of the Cruise business. It also repeated its commitment to launch its ARS business in 2019. Some have predicted that GM will provide 33,000 AV for its ARS, generating $3-$5 billion in revenue.

*Strategy*

GM has an aggressive AV strategy, as is seen in its significant investment and several critical acquisitions. It sees the importance of AVs as binary. If they aren't successful in AVs, they won't have a viable business. GM believes that it is the leader with its AV strategy, and it may very well be one of the leaders. This strategy is primarily built on a comprehensive platform that includes vertical integration of design and manufacturing, combined with the emphasis on building AVs that can be produced in volume on traditional production lines. Given the struggles that Tesla has had in getting to volume manufacturing, this may very well be an important competitive factor.

GM explicitly recognizes that autonomous rides services (ARS) is going to be the critical initial market where it must succeed and has made a bold move into this market with its Cruise business and outside investment to help fund the capital requirements.

## Ford

While following a similar general strategy as others, Ford's AV strategy has some unique elements.

The company has aggressively pursued external investments and acquisition opportunities since the latter half of 2016, backing or acquiring companies working in AI, lidar, and mapping. Its biggest autonomous technology investment came in February 2017, when it announced that it was investing $1B over the course of 5 years in AI startup Argo, and gaining a majority ownership of the company. Ford then established its autonomous vehicle efforts around Argo.

Lidar appears to be a critical technology that auto manufacturers lack, and in 2017, Ford acquired Princeton Lightwave to develop affordable lidar sensors with Princeton's help. Previously, Ford invested $75 million in Velodyne.

The importance of autonomous cars at Ford was emphasized when Ford ousted its chief executive, Mark Fields, because he failed to persuade investors and his own board that the company was moving fast enough to develop AVs. Ford's stock price dropped 40%, below Tesla's valuation. There were also rumors that former-CEO Fields failed to make a partnership with a Silicon Valley firm in 2015 that would make a huge splash and give him the investor confidence he needed. His failure to complete this left Ford struggling for an autonomous vehicle strategy that resonated with stakeholders. This upheaval at Ford may be a precursor of the challenges that lie ahead for companies that cannot adapt to that new landscape fast enough.

"Our world has changed dramatically," said William C. Ford Jr., Ford's executive chairman and great-grandson of its founder. "Look at the pace of change and the competitors coming into our space, and we need to match or beat that." The board selected Jim Hackett, an auto industry outsider, to be its next CEO because he is passionate for AVs. Some people believe that Ford is behind not only Tesla and Google but also traditional rivals like GM.

### Autonomous Ride Services

Ford refocused its strategy on developing AVs for autonomous ride service vehicles prior to selling them in retail markets.

At the end of September 2017, Ford and Lyft announced a major joint initiative. Both companies will develop software that will allow Ford's vehicles to operate with the Lyft mobile app. Lyft announced that Ford agreed to place self-driving vehicles on Lyft's open platform, which enables partners to access its one million riders per day. Ford will use this opportunity to refine its ability to connect smoothly with a ride-hailing dispatch platform. This may be indicative that Ford could supply AVs to Lyft, but it is by no means certain since Lyft has other partnerships.

In early 2015, Ford announced its "Smart Mobility Plan" to move the company forward on innovation, including vehicle connectivity and autonomous vehicles. This plan culminated in the formation of Ford Smart Mobility LLC in March 2016, a new subsidiary focused on connectivity, autonomous vehicles, and mobility, including autonomous ride services. This gives the company an option, although maybe not a good one, of introducing its own ARS.

### Delivery Vehicles

Ford has a unique focus among the other auto manufacturers on autonomous delivery vehicles. Ford and Domino's Pizza are teaming up to test AV pizza delivery in Michigan, as part of an effort to better understand how customers respond to and interact with autonomous vehicles. They are using a Ford Fusion Hybrid autonomous research vehicle, but the car won't be driving itself. Each car will be driven by a Ford safety engineer, with other researchers onboard, to study "the last 50 feet" of the customer experience.

Ford and Domino's pizza unveiled the specially-equipped Ford Fusion at the Consumer Electronics Show in Las Vegas. It comes with self-driving technology and an oven. Customers will enter a number on the touchpad near the back window and the window will then lower, revealing the pizza.

Ford also has a partnership with Postmates, the popular delivery service, to test deliveries by AVs. In the trials, Postmates' customers will be chosen at random to receive deliveries via an AV. Postmates has customers in 250 cities and more than 120,000 delivery contractors. Its appeal is offering on-demand deliveries to customers who are short on time.

### AV Development

Ford is taking an interesting strategy for the autonomous ride services (ARS) market. It is developing an entirely new AV from the ground up, instead of retro-fitting and existing model. Uniquely this new AV targeted for release in 2021 will be a hybrid, where most other companies are focusing on electric AVs. Ford's reason for hybrid is quite compelling: it wants to make this vehicle more profitable for ARS providers. A hybrid can drive much longer without refueling than electric AVs, and this can be very important to ARS providers where utilization is critical. Having a major portion of its fleet down for an hour of recharging during peak periods could become a big problem.

### Bad Weather Driving

Ford is also somewhat unique in testing AVs in less friendly environments, such as snowy Michigan, as well as in complete darkness. The company announced that it plans to roll out these highly autonomous vehicles within pre-mapped, geo-fenced areas by 2021.

### Strategy

Ford changed its strategy in the latter half of 2017. Its focus is on developing sufficiently autonomous vehicles by 2021, but only for autonomous ride services and other fleets. For the reasons previously described for ARS, autonomous ride services will reduce prices to the point where it will be cheaper to use ARS than to own a personal vehicle. It also seems that autonomous vehicles will initially be more expensive, expecting to cost tens of thousands of dollars more, and this will slow the acceptance in the retail market.

Ford's new strategy is on target. It has the best opportunity to be a leading provider of autonomous fleets to ride service companies. However, there is a lot to do to get there and many of open questions. The risk of this strategy is that the leading ARS companies may choose another manufacturer's AV or develop their own. There is a very small market of only a few ARS companies. Ford's Smart Mobility business may provide a back-up to enter ARS on its own if it doesn't get a major ARS partner, Ford may not have a significant enough position to be very successful. Its focus on developing a new hybrid AV for the ARS market may provide it with an important differentiating factor from others. I wouldn't rule out

Ford launching its own ARS though.

Early testing of AV delivery may give Ford an early advantage in the autonomous delivery market. If it does, then it could provide significant revenue as major delivery companies invest in fleets of delivery AVs.

Overall, Ford has a robust and diversified AV strategy. It has a focus on the early AV markets of ARS and AV delivery. It is trying to differentiate itself using hybrids in these markets. Then it will also actively participate in the eventual transition of the retail market to AVs.

## Fiat Chrysler

Fiat Chrysler Automobiles (FCA) designs, engineers, manufactures and sells vehicles and related parts and services, components and production systems worldwide through 162 manufacturing  facilities, 87 R&D centers, and dealers and distributors in more than 150 countries. Its stable of brands include Abarth, Alfa Romeo, Chrysler, Dodge, Fiat, Fiat Professional, Jeep, Lancia, Ram, Maserati and Mopar, the parts and service brand. FCA's CEO pushed back in a conference call following the release of the company's 2017 third-quarter earnings against a frequent assumption about the AV delays by Italian-American automaker. Instead, he said the company's approach gives FCA a chance to get a "technically correct solution at a commercially defensible price."

FCA's approach has been markedly different from that of Ford and General Motors, which are larger, have more resources and have pursued their own efforts in autonomous technology. Its CEO said it will take time, discipline and technical know-how to properly develop the technology. He offered words of caution, presumably against those who promise self-driving vehicles are right around the corner. "Don't believe the fluff. Just stick to the knitting and get to an outcome. There's no shortcut to this," Referring to the significant investment required for developing AVs, he said, "You can destroy a lot of value by chasing your tail in autonomous driving."

### AV Development

FCA's has joined a partnership with BMW, Intel and Mobileye to develop an autonomous vehicle platform. BMW and its partners Intel and Mobileye said FCA would bring engineering and other expertise to the deal, paving the way to creating an industry-wide autonomous car platform, which other carmakers could adopt. Some automakers are seeking alliances to share the high costs of developing autonomous cars. FCA Chief Executive Sergio Marchionne cited the "synergies and economies of scale" possible in joining the alliance.

### Waymo Partnership

Starting in May 2016, Fiat Chrysler has been working with Waymo to provide Pacifica Hybrid minivans for Waymo's AV development. So far it looks like FCA is simply providing the vehicles for AV development by Waymo.

FCA announced that it will provide as many as 62,000 Chrysler Pacifica Hybrid minivans to Waymo to retrofit into autonomous vehicles. It also announced that it is looking at licensing Waymo's AV technology for use in FCA-manufactured vehicles for retail customers.

Waymo's strategy is to be open to working with several automakers, they could even be interchangeable. It's possible that Waymo selected an FCA vehicle because FCA is not a viable competitor in developing AV technology.

### FCA Strategy

At this time, FCA is not a major competitor in autonomous vehicles. It is far behind in investing in the development of AVs. Its strategy in the short-term is to avoid spending money on AV development.

FCA does not have any significant advantages in its relationship with Waymo. Waymo will most likely select a car manufacturer to build its AVs, and FCA may be a viable alternative. It may end up being a subcontractor to Waymo and possibly others providing a fleet of AVs, but this is likely to be a very low margin business that could be won or lost at any time.

## Mercedes-Benz

Mercedes-Benz is part of Daimler AG, a German multi-national automotive corporation. In addition to Mercedes, Daimler owns or has shares in several car, bus, truck and motorcycle brands including Smart Automobile, Detroit Diesel, Freightliner, Western Star, and Thomas Built Buses. Daimler is the world's oldest automaker.

Mercedes-Benz

For years, it has packed its cars with semi-autonomous features, now including "Drive Pilot," which keeps the car in its lane and a safe distance from other vehicles, much like Tesla's Autopilot feature.

### AV Development

Mercedes has advanced functions in its semi-autonomous vehicles, and is continuing to introduce additional functions. The 2017 E-Class sedan has autonomous-driving capabilities that let car go farther on its own before the driver takes over and stay on track on curvier roads. It uses dual front-facing cameras to create a 3D image map of hazards up to 55 yards in front, while adaptive cruise control can now keep pace with the car ahead now at speeds up to 130 mph.

The S-Class is traditionally Mercedes-Benz's technology flagship, but the current generation was leapfrogged in some respects by the recently redesigned E-Class. With the refreshed S-Class, Mercedes plans to take it a step closer to autonomous driving. There will be a host of new and upgraded driver-assist systems in the S-Class, allowing it to operate with minimal driver input in more situations. The S-Class also gets some tech that debuted on the E-Class, including vehicle-to-vehicle (V2V) communication and remote parking. The S-Class' Distronic adaptive cruise control system will be able to automatically adjust speed through curves, intersections, and highway off ramps. This is potentially a break-through, as current production driver-assist systems aren't capable of handling much more than variations in straight-line highway driving. The new system uses map and navigation data to identify areas where the car may need to slow down. The system can also adjust speed by reading road signs or pulling speed-limit information from the navigation system. The S-Class will also get an automated lane-change function. Like Tesla's Autopilot, the system executes lane changes at the flick of a turn signal.

Mercedes' Smart Vision EQ AV is a concept AV. It is a prototype that is piloted remotely by a human outside of the vehicle. It doesn't have a steering wheel or other controls, and it only has a speed of 12 miles per hour.

### Autonomous Ride Services

In early 2017, Daimler and Uber announced that Daimler would supply AVs for use on Uber's network as the ride-hailing company works to open its platform to automakers. The arrangement calls for vehicles equipped with autonomous drive technology developed in-house by Daimler to go into service for Uber "in the coming years," the companies said in a joint statement. The substance behind this is not clear, and it may be no more than a statement of possibility.

Mercedes-Benz owner Daimler and supplier Robert Bosch are teaming up to develop self-driving cars in an alliance aimed at accelerating the production of what they call "robo-taxis" or autonomous rides services vehicles. The head of Daimler's Mercedes research and development said the he believed that self-driving taxis are the only way forward initially because privately owned cars would be too expensive in the beginning. The number of sensors, the computing power, and related systems will add tens of thousands of dollars to a car. But he believes that it will be a successful business case when you put 200, 300, or a thousand cars into service in a city. The amortization comes from not paying a driver. This is consistent with the business case for autonomous ride services described previously.

Mercedes-Benz's parent company, Daimler, later announced its intention to put self-driving taxis on the road in three years or less in a partnership with Bosch. Daimler has some experience in ridesharing with its Car2Go program and the ridesharing app it purchased in 2014, Mytaxi.

### Mercedes Strategy

Mercedes-Benz stated that its strategy involves a two-pronged approach: continue to develop the advanced driving assistance features found on the car-maker's current models, while simultaneously pursuing more advanced forms of self-driving vehicles for autonomous ride services.

The first part of that involves pushing the technologies already in use into SAE Level 3 autonomous driving. The latest S-Class is referred to as an advanced Level 2 vehicle. The progression to Level 3 isn't restricted by technology. Mercedes believes that if the restrictions on the new S-Class were lifted, it would be a Level 3 vehicle. In parallel, Mercedes is working on Level 4 and Level 5 vehicles with the most obvious use case in autonomous ride services vehicles. Mercedes believes that this enables it to offset some of the incredible costs of the project earlier in the life cycle of AVs. Mercedes also acknowledged that the high cost of sensors for fully autonomous vehicles is more feasible in autonomous ride services vehicles where the higher costs can be amortized better. Its target date for these vehicles is unclear: sometime between 2020 and 2025.

This two-path strategy to participate in both the earlier autonomous ride services market and later the AV retail market, while leveraging the significant investment in autonomous vehicles in both markets, makes sense. The risk and open question for this strategy is how will Mercedes get into the ARS market?

It may able to provide AVs to ARS companies like Uber or Lyft, but there are issues with this. Uber is developing its own AV technology and may only want to have a partner provide a "dumb vehicle". Lyft will likely work with a partner, but it could be very competitive.

Mercedes doesn't have a strong position to get people to use a ride services app that Apple, Google, Uber and Lyft have. It is too risky to do on its own, although it may try.

Overall, I see Mercedes with its high-end image able to sell AVs retail to its customer base, even at a higher price. Over time it might be able to sell almost as many AVs as it sells cars today by progressively moving its customers to increasingly autonomous capabilities.

## Audi

Audi AG is a German automobile manufacturer that designs, engineers, produces, markets and distributes luxury vehicles. It is a member of the Volkswagen Group. Audi stated that intends to be in the forefront of the autonomous-vehicle revolution. Its A8 luxury sedan, coming to the U.S. in 2018, will be the first announced car available with a SAE Level 3 autonomous-driving system.

### *AV Development*

Audi said that the A8 – its top-of-the-line large luxury sedan – was developed with autonomous driving in mind, but its new system will work only in specific, limited circumstances, at least at first.

The new A8 is the first production automobile to have been developed specially for highly automated driving. The Audi AI "traffic jam pilot" takes over driving in slow-moving traffic at up to 37 mph on highways where a physical barrier separates the two roads. Although the claimed capabilities of Audi's system are modest, the system itself does break new ground for a production vehicle as the first automaker to offer a vehicle equipped with lidar.

Audi believes its decision to limit the Level 3 driving speeds to a relatively slow speed of 37 mph was largely tied to a 99.9999% level of accuracy, a system that can be trusted in virtually every conceivable driving situation to guide the car in a way that is 99.9999% accurate. An Audi spokesperson said, commercializing a self-drive feature for speeds faster than 37 mph that was merely 99.9% accurate was unviable, even though that level of accuracy already would make it much safer than human drivers are. While the 2019 A8 will launch with the enhanced sensor package needed for these autonomous aids, the Audi AI features themselves will most likely not be available at the initial launch of the sedan. Timing will depend on where autonomous driving laws end up.

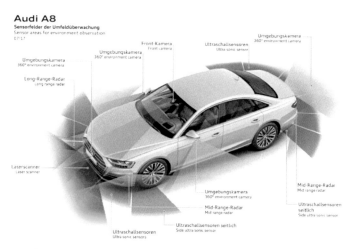

Even though others like Tesla and Mercedes have similar capabilities, Audi is the first manufacturer to promote its vehicle as a Level 3 system, claiming it does the driving for you under limited circumstances, but it needs you to take over quickly when the car's situation exceeds the system's capabilities. If the human driver is distracted, it might take several seconds to refocus, understand the car's situation, and take control. The concern is about that time lag: What's happening to the vehicle while the human is getting his or her bearings? That concern has led some automakers to declare that they would skip Level 3, and work toward a full Level 4 system that can drive the car under most circumstances instead.

Audi is also part of the German consortium — including Daimler and BMW — that bought Nokia's HERE precision mapping assets for $3.1B.

### Audi Strategy

Audi's AV strategy appears to be primarily focused on bringing AVs at retail to its customer base. Like Mercedes it may have some success in doing that even though this market will develop much slower that the ARS market.

## Volvo

The Geely-owned Volvo car brand also has made progress with self-driving passenger vehicles. With a reputation for safety innovations, Volvo labeled its autonomous vehicle endeavors "Intellisafe," with the goal of making Volvo cars "deathless" when the company fully rolls out these features to the public. For now, Volvo is planning to give 100 Swedish customers early-access to an autonomous XC90 SUV with restrictions on the area, time, and context in which the autonomous mode will be used.

The company has stated that it will accept full liability when its vehicles are in autonomous mode, and has announced plans to expand its pilot program to China and the United States. Volvo has followed rivals like BMW in setting 2021 as a target deployment date.

After searching for collaborators to work with, Volvo announced a self-driving joint venture with Swedish supplier Autoliv in January 2017. Dubbed Zenuity, the joint venture is aiming to commercialize its first driver assist systems by 2019, also making them available to other automakers.

### Autonomous Ride Services

Volvo has made good progress in its partnership with Uber. Uber is planning to buy up to 24,000 self-driving cars from Volvo in a deal potentially worth a billion dollars. Volvo will provide Uber with up to 24,000 of its flagship XC90 SUVs, equipped with autonomous technology as part of a non-exclusive deal from 2019 to 2021. Volvo will provide the vehicles, while Uber will provide the self-driving system, which is currently under development by Uber's Advanced Technologies Group. Uber has been testing prototype Volvo cars for more than a year in Arizona and Pittsburgh, with safety drivers in the front seat to intervene if the self-driving system fails.

### Volvo Strategy

Volvo appears to be comfortable with a strategy of providing a "dumb" platform for Uber to put its AV systems on. With this strategy, it will compete to provide fleet sales of vehicles to be modified by AV companies such as Uber. It's not clear now if Volvo will be building Uber's AV technology into the cars it is selling Uber or if that will be retrofit elsewhere.

It's not a bad short-term strategy while waiting for the retail AV market to mature, but it puts Volvo somewhat behind others who are being more aggressive developing their own AV technology.

## BMW

BMW has begun slowly pushing an autonomous vehicle strategy, showing off an autonomous i8 concept car and announcing a formal initiative to promote automation in its vehicles under the banner BMW iNEXT. BMW has demonstrated some strong capabilities in semi-autonomous driving, but is hesitant for fully autonomous vehicles.

The Munich-based automaker secured an alliance with Intel and Mobileye. The coalition plans to create an open standards-based platform for bringing self-driving cars to market, aiming to put vehicles on the road by 2021. BMW's Level 3 system will debut in what it calls the iNext in 2021. This is BMW's all-new vehicle architecture from the ground up - chassis, electrical systems, everything.

BMW became one of the first major carmakers to abandon its solo development of AVs in favor of teaming up with chipmaker Intel and camera and software manufacturer Mobileye to build a platform for autonomous vehicle technology by 2021. The decision followed a trip by senior executives to visit startups and suppliers to gauge BMW's competitive position. They stated that "Everybody is investing billions. Our view was that it makes sense to club together to develop some core systems as a platform."

BMW is also part of the group that bought Nokia's HERE mapping assets for $3.1B. Intel has recently taken a 15% stake in HERE as well.

### BMW Strategy

BMW, which prides itself on the driving experience, is challenged by diminishing importance this has with AVs combined with the massive investment required. It appears to be following a cautious path by relying on partnerships to minimize its investments and betting that it takes longer for AVs to have a major impact. It may also be expecting that many drivers may continue to buy BMWs for the driving experience instead of AVs for the riding experience.

## Toyota

Toyota appears to have done little in investing in autonomous driving. Its deputy chief safety technology officer publicly rejected the idea, saying "Toyota's main objective is safety, so it will not be developing a driverless car." In 2017, the Japanese automaker announced $50 million in investments over the next five years to establish research centers with both Stanford and MIT, to work on artificial intelligence and autonomous driving technology. This is an insignificant investment considering the investments others are making.

Toyota, after lagging considerably behind also poured $1B into developing AI capabilities for autonomous driving, and the first results are becoming visible, from the Verge:

> *The Toyota Research Institute (TRI) showed its first self-driving car this week, a Lexus LS 600hL test vehicle equipped with lidar, radar, and camera arrays to enable self-driving without relying too heavily on high-definition maps. The vehicle is the base for two of TRI's self-driving research paths: Chauffeur and Guardian. Chauffeur is research into Level 4 self-driving, where the car is restricted to certain geographical areas like a city or interstates, as well as Level 5 autonomy, which would work anywhere. Guardian is a driver-assist system that monitors the environment around the*

*vehicle, alerting the driver to potential hazards and stepping in to assist with crash avoidance when necessary.*

*The company is perusing both an evolutionary as well as a revolutionary strategy, but we guess that with the size and R&D budget of Toyota ($10B per year), it can afford to do that. The company has also embarked on a strategic investment in Uber.*

In exchange for an undisclosed sum Toyota will share in trials, and it will also lease cars to Uber, getting part of the ridesharing fares as payments.

### Toyota Strategy

Toyota, the largest auto manufacturer on the world in 2016, appears not to have a meaningful autonomous vehicle strategy.

## Nissan

The Chairman and CEO of Nissan and Renault promised that the group would have 10 vehicles on sale by 2020 with "significant autonomous functionality." Nissan unveiled its first public prototype in 2013 at the Nissan 360 event in California, and has since been testing an autonomous Nissan LEAF

on the roads of Tokyo. Nissan and Toyota also announced a joint effort to develop standardized "intelligent" maps.

Nissan introduced its new ProPilot system, a suite of sensors, controllers, actuators and systems in the 2018 Nissan Leaf electric vehicle. It's classified as an SAE Level 2 system, which means that a human being still monitors the system, still guides it, and still is ultimately responsible for what the car does. The steering, braking and acceleration are managed autonomously by the car, which will operate those mechanisms under specific driving conditions, on the highway where clear lane markers are present. Nissan refers to this as "driver in the loop," meaning that the person is still there and still driving the car. Unlike systems from Mercedes-Benz or Tesla (or the upcoming Super Cruise feature from Cadillac), drivers cannot take their hands off the steering wheel.

### Nissan Strategy

Nissan appears to be a follower, rather than a leader, in autonomous vehicles, but it doesn't want to be left out completely. It is not investing as much as some other car manufacturers in autonomous systems, and it will most likely focus on a Tesla-like vehicle with its Leaf, but will be more semi-autonomous.

## Tier-One Suppliers

Manufacturers sometimes refer to companies in their supply chain as tier-one and tier-two suppliers. The terms indicate the commercial distance in the relationship between the manufacturer and supplier. Although supply tiers can apply to any industry, the terms most commonly describe manufacturer and supplier relationships in the automotive industry.

An original equipment manufacturer, or OEM, refers to a company that makes a final product for the consumer marketplace, for example, Ford and General Motors. Tier-one companies are direct suppliers to OEMs. The term is especially common in the automobile industry and refers to major suppliers of parts to OEMs. Tier-two companies are the key suppliers to tier-one suppliers, without supplying a product directly to OEM companies. A single company could be a tier-one supplier for one product and a tier-two supplier for a different product line.

Some of the tier-one suppliers may play an important role in autonomous vehicles, but it is too early to predict specific strategies and the chances of success, nevertheless, here are the AV activities at several of the more important ones.

## Bosch

Bosch, the largest auto industry supplier, claims to be investing heavily in technology to supply AVs. Bosch has partnered with Nvidia on AV computing, with TomTom on mapping systems, and with Mercedes on automated valet and taxi concepts. Bosch's concept car featured a screen-covered dashboard. Of all the tier-one suppliers, Bosch is perhaps best positioned to produce autonomous vehicle technology given its size and scope. It covers many aspects of AVs, including the sensors (except lidar) and the electronics. It also has system competencies.

It announced a $1.1 billion facility that will produce semiconductors for AVs, as well as for other markets such as smart homes and smart city infrastructure. The new Dresden-based chip fab is set to start producing silicon commercially in 2021.

## Continental

Continental made some AV investments, including a minority stake in French autonomous driving company EasyMile. Continental's CUbE test vehicle is based on an EasyMile shuttle. In 2016, Continental acquired the lidar business of California company Advanced Scientific Concepts Inc. It also has a concept car demonstrating technology called the "Cruising Chauffeur" that can automatically accelerate, brake and change lanes at highway speeds.

Overall it seems like it is taking a more gradual route, focusing on technologies that still require the driver to take over in certain cases.

## Aptiv (Delphi)

Delphi Automotive split itself into two with a new company, Aptiv, focusing primarily on the advanced technologies of semi-autonomous vehicles and AVs. It designs and manufactures vehicle components, and provides electrical and electronic and active safety technology solutions. Its business segments include electrical/electronic architecture, and electronics & safety. The electrical/electronic architecture segment provides complete design of the vehicle's electrical architecture, including connectors, wiring assemblies. Aptiv claims to have a focus on cutting the cost of autonomous technology by more than 90% by 2025.

Delphi, now Aptiv, purchased Nutonomy, a Boston-based company that develops autonomous vehicle technology for $450 million. With this this purchase, Aptiv hopes to accelerate the pace of developing AVs. Aptiv sees the initial AV market for commercial vehicles, with the technology then bleeding over into consumer vehicles over time. It has a pilot test vehicle that's based on a BMW 5-Series. The tech package on this vehicle includes more sensors than previously, but it offers a design that's more integrated into the existing lines of the consumer vehicle. Aptiv's plan is to continue to move towards something that's more production-ready, both in terms of capabilities, and in terms of meeting the design expectations of manufacturers and consumers.

### Magna

Magna International Inc. (Magna) is a global automotive supplier. As a tier-one supplier, it offers a broad range of product capabilities include producing body, chassis, exterior, seating, powertrain, electronic, active driver assistance, vision, closure, and roof systems and modules, as well as vehicle engineering and contract manufacturing.

Magna joined with BMW Group, Intel and Mobileye as a tier-one technology integrator to help industrialize and customize a domain controller. Magna claims to have identified some of the world's most advanced technologies and 'auto-qualified' them for use in the auto industry. For example, its Icon Radar takes the best of military technology and improves on it for automotive use, taking a significant step forward toward full autonomy."

Perhaps most importantly, Magna International as the world's largest contract manufacturer of vehicles, could be perfect as Apple's partner to manufacture its ARS vehicle. This would be like Foxconn, the iPhone manufacturing partner for Apple. In 2016 a German publication created a rumor that Apple had a secret lab in Berlin, suggesting its sources believed Apple was looking at a Magna factory in Austria. In addition, Bloomberg reported that a team of Magna engineers were working with Apple in its Sunnyvale car offices.

# AV Computing Platform

The computer platform powering AVs is perhaps the most important new component. This is what interprets sensor data, executes the artificial vehicle intelligence software, and directs the vehicle's movements. Currently, there are two primary companies in the market: Intel and Nvidia. AMD is also a potential competitor in this market. In addition, some companies like Tesla and Apple may be developing their own proprietary computing platform. Of the two primary companies, Nvidia currently has a big lead over Intel.

For a computing platform, design wins are critical. This means that the AV manufacturer designs its AV around that computing platform, locking in sales for the entire generation of that AV, and most likely all the AVs made by that manufacturer for the foreseeable future. The computing platform could be the most expensive component of an AV with estimates of $600-$1,000 per vehicle. Some estimate this market could be as much as $10 billion, including semi-autonomous and autonomous vehicles.

## Nvidia

Since 2014, Nvidia has shifted to become a platform company focused on four markets – gaming, professional visualization, data centers and auto. Auto is only about 8% of its revenue but expected to grow rapidly. Nvidia's deep expertise in graphics processing gives it an advantage in this new market.

Nvidia created one of the first computing platforms for AVs with its Drive PX2 platform. Since then it has continued to advance this technology and sign up many potential customers.

### Drive PX2 Platform

Nvidia began its entry into AV computing in the spring of 2016 when it started shipping its Drive PX 2 AI (artificial intelligence) car platform. This is a supercomputer for processing and interpreting all the data taken in by the various sensors about the surroundings of semi-autonomous and fully-autonomous vehicles.

Most importantly, Nvidia integrated this advanced technology into a common platform. It provides the power and flexibility to develop and deploy artificial intelligence (AI) systems for autonomous vehicles. Its end-to-end approach leverages Nvidia DriveWorks software and allows vehicles to receive over-the-air updates to add new features and capabilities throughout the life of a vehicle.

Drive PX can understand in real-time what's happening around the vehicle, precisely locate itself on an HD map, and plan a safe path forward. It's an advanced autonomous vehicle platform that combines deep learning, sensor fusion, and surround vision. Drive PX systems can fuse data from multiple cameras, as well as lidar, radar, and ultrasonic sensors. This allows algorithms to accurately understand the full 360-degree environment around the car to produce a robust representation, including static and dynamic objects. Use of deep neural networks (DNN) for the detection and classification of objects dramatically increases the accuracy of the resulting fused sensor data.

As a platform, it provides a scalable architecture that is available in a variety of configurations. These range from one passively cooled mobile processor operating at 10 watts, to a multi-chip configuration with two mobile processors and two discrete graphic processing units (GPUs) delivering 24 trillion deep learning operations per second. Multiple Drive PX platforms can be used in parallel to enable fully autonomous driving.

In October 2017, Nvidia introduced the Pegasus version of Drive PX, which it plans to ship in the second half of 2018. It can handle 320 trillion operations per second (TOPS). It specializes in deep learning operations that Pegasus is especially good at, rather than general-purpose computing power. Pegasus' performance easily outclasses the 24 TOPS delivered by the Drive PX 2, which

Nvidia unveiled in early 2016. It also easily surpasses the 30 TOPS promised by the Drive PX Xavier, a smaller, less power-hungry board set to ship in Q1 2018.

Nvidia is aggressively designing and releasing new Drive PX variations, with each new model significantly more powerful or less power-hungry than its predecessor.

In tandem with the Pegasus announcement, Nvidia extended its platform with Drive IX, a software development kit (SDK) for creating AI Co-Pilot-powered systems. The Nvidia DriveWorks SDK contains reference applications, tools and library modules. It also includes a run-time pipeline framework that integrates every aspect of the driving pipeline, from detection to mapping and localization to path planning to visualization. It also showed off a simulator that uses eight of Nvidia's DGX-1 servers (they contain powerful Nvidia server GPUs) to simulate 300,000 miles of driving within five hours.

At this point, Nvidia appears to be the clear leader in the technology in computer processing for autonomous vehicles.

### Partners (Customers)

Nvidia has more than 225 auto manufacturers, truck manufacturers, suppliers, research organizations, and start-ups as partners and potential customers developing AVs using the Drive PX2 platform. Some of these use the platform in vehicles that are already on the market or are slated to come to market within the next five years:

- *Tesla*: Tesla uses the Drive PX 2 to power its Autopilot on all new Model S and Model X vehicles, and its lower-priced Model 3.
- *Volkswagen's Audi*: Nvidia and Audi announced their collaboration to put advanced AI cars on the road starting in 2020. Audi unveiled its new 2018 A8, which it claims is the world's first Level 3 autonomous car to go into production.
- *Daimler's Mercedes-Benz*: A partnership to bring a Nvidia AI-powered Mercedes car to market has been announced.
- *Toyota*: The giant automaker will use the platform to power its autonomous driving systems in vehicles planned for market introduction within the next five years.
- *Volvo*: The Swedish auto maker and Swedish auto supplier Autoliv announced they're teaming with Nvidia to make production vehicles built on the Drive PX 2 platform by 2021.
- *Baidu*: The Chinese search engine giant announced that it is adopting Drive PX 2 for its autonomous vehicle initiative and will also use the platform in developing self-driving cars with major Chinese automakers.

### Nvidia Strategy

Nvidia has a very aggressive strategy to be the leader in providing computing platforms to AV manufacturers. It leveraged its deep experience with graphics processing to be the innovator in this potentially major market. It aggressively created a complete platform with multiple product variations from that platform and a broad range of software and development tools, so customers can more easily develop products from the platform. Then it just introduced an incredibly powerful high-end product from this platform that will be available just at the right time for the ramp up of AVs. Overall, this creates a powerful strategy that should enable Nvidia to be the leader in this market.

## Intel (Mobileye)

Intel got into the autonomous vehicle market relatively late, but it is trying to make up for lost time. The chipmaker plans on investing $250 million over a two-year period to develop self-driving technology, and in 2017 it acquired Jerusalem-based auto-vision company Mobileye for $15 billion, a company that generated less than $360 million in revenue in 2016. In buying Mobileye, Intel inherited the company's existing relationships with 25 automotive partners, including tier-one automotive suppliers.

### Mobileye

Mobileye is a leading supplier of software that enables Advanced Driver Assist Systems (ADAS), with more than 25 automaker partners including some of the world's largest. Mobileye's strategy is primarily based on a single-lens camera (mono-camera) as the primary sensor to support Advanced Driver Assistance Systems and eventually autonomous vehicles. Mobileye's vision technology for ADAS is deployed on over 15 million vehicles.

In a nutshell, Mobileye makes computer systems that can parse images from cameras, designed for use in automobiles and trucks. Mobileye's systems, which combine hardware and software, are already in many of the advanced driver-assist systems (ADAS) that have become common in new cars, including features like adaptive cruise control, lane-departure warnings, and parking-assist systems.

### EyeQ Processor

With the acquisition of Mobileye, Intel is focusing on the EyeQ family of processors for autonomous vehicles. What sets the EyeQ family of system-on-chip (SoC) devices apart from the competition is EyeQ's ability to support complex and computationally intense vision processing, and still maintain low power consumption even while located on the windshield. Mobileye achieves the power-performance-cost targets by employing proprietary computation cores (known as accelerators), which are optimized for a wide variety of computer-vision, signal-processing, and machine-learning tasks, including deep neural networks. These accelerator cores have been designed specifically to address the needs of ADAS and autonomous driving.

The high-end device from this family is the EyeQ5, which can process 24 trillion operations per second, and notably will consume only 10 watts of electricity. The EyeQ5 will be the Mobileye's first product aimed at fully autonomous vehicles, which are classified as SAE automation levels 4 and 5. Production is anticipated for 2020 which means it may be in cars beginning in 2022. The EyeQ5 was first announced by Mobileye in 2016, prior to the acquisition by Intel, but some of the specifications for the product have since changed based on Intel's expectations and product simulations.

Mobileye continues its long-standing cooperation with STMicroelectronics, leveraging STM's substantial experience in automotive-grade designs. STM provides support in state-of-the-art physical implementation, as well as automotive-grade memories, high-speed interfaces, and system-in-package design.

### Partnerships

Intel also acquired a 15 percent ownership stake in HERE, a global provider of digital maps and location-based services. The companies plan to jointly develop a highly scalable proof-of-concept architecture that supports real-time updates of high-definition (HD) maps for highly and fully automated driving.

### Intel Strategy

Intel waited too long to invest in computing platforms for AVs. Its strategy is based on acquiring Mobileye, which gives it some critical technology and a relationship with automotive companies. It looks like the EyeQ product line will lead its strategy for first-generation computing. Intel may be late for the first generation of AVs, but it is a strong aggressive company that could catch up.

## Other Electronic Components

In addition to and supplementing computer processing, there will be a wide range of components used in AVs. It's too early to identify these strategies since the systems architecture for AVs is still in process. This architecture will determine what additional electronic components will be necessary, and which ones will be imbedded in the primary computing platform. Here are some highlights of the more notable ones.

### NXP

NXP has a broad portfolio of products meeting automotive-grade functional-safety requirements. It is well positioned for AVs since it already supplies Advanced Driver Assistance Systems (ADAS) processors to eight of the top 10 carmakers.

The NXP's BlueBox engine pulls together input from radar, lidar (laser), and vision sensors, and it has secure V2X communications built in. In an AV system, multiple streams of sensor data could be routed to the BlueBox engine and combined to create a complete 360° situational real-time model of the physical environment around the vehicle. The BlueBox engine is a Linux-based open platform that lets carmakers or third-party software vendors build their own autonomous car applications. The NXP S32V processor takes the sensor inputs from the platform's NXP silicon-powered lidar, radar and vision nodes, and creates a map via its sensor fusion capabilities.

NXPs strategy appears to leverage its broad application of electronic devices into a more comprehensive platform for AVs, particularly doing sensor fusion. Most likely its components will be designed into many AVs, but it's unclear if it will become the primary computing platform.

## AMD

The graphics chip maker has launched AMD Radeon Instinct, a combination of hardware and open-source software which it hopes will make its graphics processing unit (GPU) accelerators more broadly adopted in machine intelligence. While it may be lagging Nvidia in tackling the segment, AMD specializes in both CPU and GPU design, so it has the potential to compete on both ends of the spectrum.

Tesla has started working on developing its own processing capability in conjunction with AMD.

### Other Component Manufacturers

The incumbent chip makers that already supply chips to the auto industry for ADAS are also expected to continue to develop more chips for AVs as well. These companies include Qualcomm, which develops modem chips for infotainment systems, and is working on a connected car platform; Cypress Semiconductor Inc. which supplies memory, wireless radio chips, power management and other chips to car makers; On Semiconductor Corp. which develops image sensors for parking assistance and surround/rear view; Xilinx Inc. which provides embedded chips for cameras and vision systems; Texas Instruments, a supplier for a range of infotainment and ADAS functions including a driver assistance system-on-a-chip; and Maxim Integrated Products Inc., which develops chips for infotainment systems, electric-car battery management, and high-speed communications chips for vehicle video cameras.

# Lidar Companies

Lidar is expected to be one of the primary new technologies that make AVs feasible, despite Tesla designing its cars without lidar. Since this is a new technology, it's difficult to project who will be successful, other than Velodyne, which already has a significant lead.

Private and corporate investors have poured more than $700 million into startups developing lidar systems for self-driving cars, according to a Reuters analysis of publicly available investment data.

Princeton Lightwave is one of at least four of those startups to be acquired in the past two years. Others include Tyto Lidar, a small San Francisco Bay area company purchased by Uber Technologies in 2016, and Pasadena, California-based Strobe Inc., which General Motors bought in 2017.

### Velodyne

Velodyne is a U.S. based group of three subsidiary companies: Velodyne Acoustics, Velodyne Marine and Velodyne Lidar. Co-Founded by David and Bruce Hall in 1983, it started as an audio company. Then it discovered a niche for

lidar sensor technology while participating in a competition to design an autonomous vehicle prototype based on stereovision technology. Velodyne's early
success in AVs has made it the leading manufacturer of advanced lidar sensors
with a 360-degree environmental view at a range of about 100 meters, providing
3D mapping. Velodyne has a large customer base for AVs.

Originally priced at about $8,000, the VLP-16 Puck was one of the most
expensive components for AVs. The HDL-64 uses four times the number of lasers, and is much more precise and expensive. At the start of 2018, Velodyne cut
the price of the VLP-16 Puck in half due to manufacturing economies of scale.
Further prices reductions are expected.

### LeddarTech

LeddarTech, established in 2007, is an offshoot of Canada's optics and pho-
tonic research institute (INO). Its products range from LeddarCore ICs, M16
Sensor modules to D-tech traffic sensors. Several automotive OEMs have incor-
porated LeddarCore ICs to develop lidar solutions with customized optical
configurations.

### Innoviz

Innoviz, an Israel-based startup, develops lidar sensor technology to provide
a lidar solution to eminent automotive OEMs. Magna has collaborated with Innoviz to utilize its high-definition solid-state lidar (HD-SSL) sensors to provide a
complete sensor-solution package for self-driving vehicles to the automotive manufacturers.

### TetraVue

TetraVue, founded in the USA in 2008, provides the TetraVue solid state
HD Lidar, which is long-range, high-resolution, and low-power consuming.

### Luminar

Luminar, founded by Austin Russell in 2012, is trying to create a new type
of lidar for autonomous vehicles. Russell's passion for creating an advanced lidar
sensor was fueled by his Theil Fellowship at an age of 18. He founded Luminar
when he turned 22. It has a uniquely innovative product where everything from
creating its own chip to circuit design is done by the company. Its product claims
to have a longer range and a higher resolution, which gives the vehicle an optimal
reaction time to avoid accidents.

The company has been selected for testing by four major autonomous vehicle programs and is now in a 10,000-unit production run. With its new 50,000-sq-
ft manufacturing facility in Orlando, FL, it expects to make deliveries to its strategic partners later this year.

## Oryx

Oryx's flash automotive lidar achieves the depth of vision performance required for true autonomous driving – yet it's as simple and robust as a digital

camera. A coherent, frequency-modulated lidar, it has a million times better signal-to-noise ratio than scanning systems, is not blinded by the sun or other lidars, and produces both range and velocity data for each point in its field of view. Solid-state, with no mechanics at any scale, enables auto-grade durability.

## Investment Recommendations

In two of my previous books, *Product Strategy for High-Technology Companies*, the first edition in 1994 and the second edition in 2000, some readers found the examples turned into very good investments. In particular, my examples of Intel, Microsoft, and Compaq in 1994 as companies that could leverage a platform strategy for exceptional growth turned into good investments at the time. The 2000 edition introduced the exciting prospect of Amazon's long-term growth possibilities.

There is much more to investing than simply understanding a company's strategy. Valuation, timing, risk, and many more factors complicate investment decisions. Many of the companies that I cite here have complications. Apple and Google are already very large and cannot be considered pure play investments. Uber and Lyft are private companies. The auto manufacturers will have more downside that upside in the transformation.

So, two important points. First, I have an equity position – individually, through our family trust or our family charitable foundation – in many of the companies that I write about in this chapter, and I am most likely biased. Second, **this is not intended to be investment advice so please don't use it as such.**

# Chapter 8
# Disruptions Caused by Autonomous Vehicles

Autonomous Vehicles (AVs) are an extremely disruptive innovation. In general, disruptive innovation refers to an innovation that creates a new market and value network, which eventually disrupts an existing market and value network, typically displacing established market leading firms. The term was defined and first analyzed by Harvard Business School professor Clayton M. Christensen. Clay Christensen's primary focus was on innovative disruption to particular companies, but he also applied it more broadly to business sectors.

There are degrees of disruptive innovation. Smartphones, for example, were not a significant disruption. Cell phones to some degree disrupted land-line phone services, but smartphones were an innovation that added vast new capabilities that weren't previously available. Also to some extent, the original automobile wasn't very disruptive because it provided transportation to most people that was not already available. It did displace stagecoaches and horses to some degree, but it wasn't terribly disruptive.

Autonomous vehicles will create an extreme degree of disruption. Perhaps, more than any other disruptive innovation in history. The primary reason behind this is that it will displace a huge existing industry, transportation, along with all its supporting industries. When we look at the enormous benefits of AVs regarding the virtual elimination of auto accidents, a significant reduction in the cost of transportation, etc. we need to realize that there is a corresponding offset. In other words, these savings need to come from someplace.

In many cases, these disruptions don't come just from AVs replacing current vehicles; they come from the increased efficiency and lower transportation cost of Autonomous Ride Services (ARS). Simply stated: if vehicles currently sit idle 95% of

*The overall loss of jobs will be massive.*

the time and ARS vehicles are utilized 40%-60% of the time, then we will need fewer vehicles. Again, simple math dictates that if a new multi-hundred billion-dollar, possibly more than a trillion-dollar, industry (Autonomous Ride Services) is created, and the cost of transportation is significantly reduced, then there needs to be some enormous disruption to current transportation industries.

These disruptions will cause significant loss of jobs. Sure, there will be some new jobs created, but the overall net loss of jobs will be massive.

I'll analyze predicted disruptions to the current transportation industry, as well as some of the related service industries. In doing this analysis, I realized that the disruptions will be different in different industries and will be caused by different factors.

## Disruptions to Auto Manufacturing

Let's start with some basic facts. There are approximately 250 million registered vehicles (excluding trucks) in the United States and around 17 million new cars sold annually. Globally there are roughly a billion vehicles with nearly 60 million new cars produced annually. Approximately 3 trillion miles are driven in the United States annually.

The most significant disruption to auto manufacturing will come from ARS. ARS will provide more transportation per vehicle than current cars that sit idle 95% of the time. Let's assume that an autonomous rides services (ARS) vehicle is utilized 40%-50% of the time. Let's also assume, for the sake of simplicity, that lower wasted travel offsets the increase in miles traveled because of convenience and lower cost. Every ARS vehicle will displace 8-10 individually-owned vehicles. For example, a million ARS vehicles will replace the need for 8-10 million individually-owned vehicles.

As I stated earlier, ARS will become a significant new industry in the United States and most developed countries. The extent of the disruption from ARS will depend on what percentage of miles driven shifts from individually-owned vehicles to ARS. For the sake of illustration, let's create one potential scenario and assume that this shift is 30% at some time in the future. This would translate into 30% fewer individually-owned cars or approximately 60 million fewer cars. As a note, this would require approximately 6 - 7.5 million ARS vehicles, given a displacement ratio of 8-10 to 1.

### New Car Manufacturers

This decline in individually-owned cars would significantly impact new car manufacturers over time, but not over a long period of time. Assuming 10-12-year average car ownership, the decline would eventually be approximately 6 million cars annually from 17 million. This 30%, or higher, drop in new car sales would create an enormous disruption to the auto industry, both in

*A 30% or even greater reduction in new car sales would create enormous disruption in the auto industry.*

142

the United States and globally.

The auto industry relies on manufacturing economies of scale, and any reduction to this extent would threaten the viability of most manufacturing. I expect that there would be a significant reduction in auto brands and considerable consolidation of companies. These estimates are for the United States alone. On a global basis, the reduction to the 60 million cars manufactured annually would be much higher.

Of course, these estimates rely on some assumptions, particularly the penetration rate of ARS. If the penetration is only half of the 30% in the base scenario, then the reduction would just be 3 million, but still a big reduction. If it were twice that estimate, then the reduction would be closer to 12 million cars annually.

The other factor determining the impact on new car manufacturing is how fast this reduction comes from new cars instead of used cars. The conservative assumption is that it would be proportional. Most likely though, people will stop buying new cars faster, so the impact would come more quickly to new car manufacturers. Perhaps 10 million or more per year.

For new car manufacturers, fleet sales of ARS vehicles may be a source of new sales for a few, but remember the displacement ratio of 8–10 to 1, so it won't replace the loss, and there will also be a reduction in other fleet sales, especially those to rental car companies.

More than 900,000 people work in automobile manufacturers in the United States, but since this is a such a global industry, it's impossible to predict how many job reductions will occur in this country.

Most auto manufacturers recognize this coming disruption, which is why they are jockeying to have a leading role in AVs.

## Used Car Market

Autonomous vehicles will also disrupt the used-car market. According to Edmunds, almost 40 million used cars are sold annually out of the approximately 250 million cars in use. I expect that AVs and ARS will significantly reduce the demand for used cars and depress pricing. The overall reduction in individually-owned cars will reduce the demand for used cars as much as new cars. Additionally, cars without at least semi-autonomous capabilities will have much less value than they have today. This reduction will create an excess of supply over demand, driving prices lower.

The reductions in demand and value have several other implications as well. Lower resale values will force some to hold onto their cars longer when trade-in values decline. The reduction in resale value will also have an impact on leasing, where the expected residual values will be much lower. As an additional note, the residual values for AVs will also be lower because of the expected rapid change in technology.

# Disruption to Supporting Industries

Many industries support the auto industry, and there will also be a disruption in most of these. These industries include auto dealerships, auto repair shops, gas stations, auto insurance agents and companies, lease financing companies, retail auto parts suppliers, driver training, and many others. Let's look more closely at some of these.

## Auto Dealers

A 30%, or higher, decline in new car sales will significantly disrupt auto dealerships. Even a 20% decline will affect the viability of many dealers who rely on a minimum volume of sales to cover operating costs such as rent and overhead. I expect that many dealerships will consolidate or just go out of business until there is a new equilibrium with fewer dealers.

Also, the reduction in residual values may delay new car sales, and the decrease in individually-owned cars will reduce service revenue.

Some select dealerships will shift to focus on selling autonomous vehicles. Doing this will require some significant investments and retraining but will provide them with future opportunities. Since dealerships are based on a franchise to sell specific brands, it will be interesting to see if the auto manufacturers create new brands for AVs and try to sign-up new dealerships.

Approximately 1.3 million people work for automobile dealers in the United States.

## Gas Stations

There are approximately 150,000 gas stations in the United States, but this number is declining already. Fewer gas stations will be needed because of the decline in the number of individually-owned vehicles, but even more broadly the demand for gasoline sales will be reduced by the switch to electric, and hybrid vehicles. The increase in autonomous vehicles will accelerate that trend. Within a decade, the demand at the gas pump will significantly diminish.

Even a 25% reduction in individually-owned cars will drive a significant reduction in volume at gas stations. There won't be any offsetting business from ARS vehicles because they will be mostly electric (or hybrid) and will be fleets, so that they won't use gas sta-

*By 2035, gas stations as we know them today will be virtually extinct.*

tions. Over time, I expect that almost all vehicles will be autonomous and mostly electric, so gas stations as we know them today will mostly disappear. Some new gas stations will provide electric charging capabilities to offset the loss of gasoline sales. Stations like many of the RaceTrac locations have additional space for charging stations where multiple vehicles could sit for an hour or more, and they offer food service and tables for customers to eat while waiting for their vehicle to charge. However, unlike gasoline, electric vehicles can be charged at home.

It's likely that by 2035, gas stations as we know them today will be virtually extinct.

## Auto Insurance Companies

Insurers such as State Farm Insurance, Allstate Corp., Liberty Mutual Group, GEICO, Citigroup Inc., and Travelers Group could lose the majority of the $200 billion in personal auto premiums. Initially, the expected decline in auto accidents from AVs may benefit their expense for claims, but this will quickly be reflected in lower premiums and revenues.

*Auto insurance premiums will be reduced by 50%, and possibly more.*

On top of this, ARS will reduce the number individually-owned vehicles, further eroding the market. ARS vehicles will be primarily in fleets with lower premiums, and many of these may be self-insured by the companies operating the fleets.

The impact on auto insurance companies will be severe. Auto insurance premiums will be reduced by 50%, and possibly a lot more.

## Insurance Agents

Corresponding to the reduction in auto insurance, those agencies that rely on selling auto insurance policies will be significantly affected. The reduction of 50% or more in insurance premiums will substantially reduce property and casualty insurance premium revenues.

About 35,000 insurance agencies employ approximately 1 million people in the United States. Auto insurance makes up about 40% of the property and casualty sector of the insurance industry and about 15%-20% of the total insurance industry of more than $1 trillion. The anticipated reduction in auto insurance premiums will probably drive consolidation in the number of agencies.

## Auto Repair Shops

There are approximately 225,000 auto repair facilities in the United States, employing nearly 850,00 people. Many of these are independent auto repair shops. The reduction in the number of vehicles will significantly reduce this business, but this will be amplified. AVs will be too sophisticated for most of these to repair. Also, as the value of used cars declines, fewer will be repaired.

A 50% reduction over time is probably conservative.

## Retail Auto Part Suppliers

Auto part companies that primarily sell retail may see a massive drop in business as car ownership goes down and autonomous ride services maintain their fleets. Approximately 500,000 people work in retail auto parts stores and related businesses, such as tire stores.

## Driver Education Companies

More people of all ages, especially young people, aren't getting their driver's license, according to a study by the University of Michigan Transportation Research Institute. The percentage of people aged 16 to 44 with driver's licenses in 1983, 2008, 2011 and 2014 declined in each period. The need for driver education will reduce much further with ARS and AVs, as more new driving-age teenagers realize that they won't need a license. By some reports, driver education teachers can make $60,000 to as much as $100,000 per year. In some states, driver's education is state funded.

Correspondingly, there will be a reduction in the need for state motor vehicle registrations and licenses. Driver's licenses will slowly disappear, as will the Department of Motor Vehicles in most states. Other forms of ID may emerge as people no longer carry driver's licenses

### Auto Financing

With lower automobile ownership because of ARS, there will be a reduction in auto financing and leasing. The jobs of people who arrange the leases and financing will be greatly reduced, as will the amount of capital invested in auto financing.

## Taxi and Ridesharing

Eventually, ARS will replace taxis. For a while, there will still be some taxis driven by human taxi drivers, but this will decline to the point where taxi operations will be unprofitable. There are approximately 200,000 taxi drivers in the United States (depending on the estimate). While not a high paying job, for many drivers it is their primary source of income. *Millions of ridesharing drivers will lose their jobs.* Even worse, many drivers in large cities invested a lot of money for their taxi medallions giving them the right to have a taxi. Some paid as much as $500,000 to $800,000 and will be bankrupt with the elimination of the taxi industry.

There are another estimated 50,000-100,00 limo drivers in the United States with many of those working part-time. ARS will also displace most of these jobs.

The number of ridesharing drivers is exceptionally high. Uber claimed to have 2 million drivers worldwide in 2017, and the number was growing rapidly. In the United States, its estimate is a million drivers. This would place the total estimate for ridesharing drivers at approximately 4-5 million worldwide and about 1.5 million in the U.S. Many of these drivers are part-time. Ridesharing is already replacing taxis, but autonomous ride services will replace ridesharing. Initially, ARS will coexist with ridesharing by providing the capacity for ride service growth in some metropolitan areas. In other areas that don't have ARS, ridesharing will continue to grow. But inevitably, the economic and convenience advantages of ARS will completely replace ridesharing, eliminating millions of jobs.

# Truck Drivers

There are approximately 3.5 million professional truck drivers in the United States, according to estimates by the American Trucking Association. The total number of people employed in the industry, including those in positions that do not entail driving, exceeds 8.7 million. At the current time, the need for truck drivers is increasing by an estimated 100,000 per year, and it is challenging to meet that increasing demand.

*Autonomous trucks will reduce the need for many of the 3.5 million truck drivers.*

Autonomous trucks will begin slowly replace drivers. At first, it will supplement drivers still having drivers in the truck, but allowing them to rest and sleep when they are driving autonomously. This productivity improvement will initially eliminate the need for more drivers. Eventually, autonomous trucks will begin to reduce the need for 3.5 million truck drivers.

# Rental Car Companies

The rental car industry will be upended as autonomous ride services become cheaper and ubiquitous. Why would a consumer rent a car if the cost of ordering a ride service from Uber or Apple was 80%-90% cheaper? Car rental is a $75 billion industry that will need to change as it shrinks. A few of these companies may be able to innovate to survive.

JP Morgan analysts stated a belief in 2017 that the advent of fully autonomous driving will level the playing field for rental car companies and rideshare solutions, making them the same, additionally necessitating investments in fleet management services by rideshare companies. There is some rotation from car rentals toward ridesharing for short rental periods with low utilization. Both Hertz and Avis estimate that such substitution is feasible only in less than 10% of their transactions, but the use of technology to further lower cost per mile and increased convenience will increasingly cause overlap and competition for the same customer between ARS and car rental businesses.

They argue that while the primary cost associated with ridesharing is the labor cost of the driver and autonomous ride services will eliminate this expense. Then these companies will need to own their fleet of vehicles. They would need to build out expensive fleet management infrastructure to maintain, repair and clean vehicles, and to be able to sell vehicles at the end of the holding period, similar to traditional rental car companies today.

I disagree with the assessment that rental car companies will be successful participants in autonomous ride services. It's too far-fetched for rental car companies to develop autonomous cars, but they are experts at managing and maintaining large fleets of cars spread out across a variety of geographies. They have a model for buying cars in volume, managing them, and then selling them from their fleets. They also can service and maintain large fleets, which includes a wide geographical footprint. A few of the rental car companies may be able to transform themselves by servicing autonomous fleets of cars for ARS companies such as Apple, Waymo, Uber, Lyft, etc. But they won't control the ride services businesses; it will be more like a low-margin building maintenance business.

There also may be an opportunity for car rental companies to provide autonomous vehicles for medium-distance rentals, say from Boston to New York. They have facilities throughout the United States so they can offer pick-up in one location and drop-off in another. ARS will primarily be focused on fleets based in particular metropolitan areas. But this opportunity still remains to be seen.

*Auto rental is a large industry that is at risk.*

Auto rental is a large industry that is at risk. Enterprise has $14 billion in revenue. Hertz has revenue of almost $9 billion. Avis has $6 billion in revenue. A couple of these companies are trying to participate in ARS.

### Avis

Waymo plans to offer an autonomous ride-hailing service in Phoenix, Arizona, using a fleet of adapted Chrysler Pacifica hybrid minivans with its self-driving technology. The company announced a deal with Avis to service and store the vehicles. Waymo will own the autonomous test fleet and will pay Avis to look after the vehicles.

Under a multi-year agreement, Avis will service and store 600 of Waymo's self-driving cars. The arrangement solves a fundamental problem faced by companies working to incorporate autonomous vehicles into ridesharing services. The vehicles need standard maintenance and cleaning, so they are ready for passengers. While it's too early to project that this will be a broader partnership, it is the start of an interesting relationship that may, in fact, blossom into something much bigger.

### Hertz

Apple is leasing a small fleet of cars from Hertz Global Holdings Inc. to test autonomous technology. This agreement may be part of a more significant deal. It is leasing Lexus RX450h sport-utility vehicles from Hertz's Donlen fleet-management unit, according to documents released by the California Department of Motor Vehicles. At the beginning of 2018, Apple increased its autonomous fleet by 24 more vehicles, and then to 62 in total.

## Parking Lots

There are more than a billion parking spaces in the United States. That's correct, a billion. And it's a $100 billion industry. The parking industry segments into on-street parking and off-street parking. On-street parking (i.e., parking meters), represents about one-third of all parking-related revenue in the U.S. and is typically controlled by cities and municipalities. Off-street parking (i.e., garages and surface lots), which represents about two-thirds of all parking-related revenue in the U.S., is primarily owned by private enterprise.

In the U.S., there are more than 40,000 garages and surface parking lots. Owners of these facilities rarely manage them, instead relying on parking operators and equipment providers (that provide access and revenue control solutions) to maximize parking revenue. The number of parking lot attendants is probably well over 100,000.

While ARS won't eliminate the need for parking lots, it may significantly reduce that need. In many cities, a significant reduction in parking will make garages and lots unprofitable and force them to close.

## Short/Medium-Distance Air Travel

Disruptions will not be limited to auto industries. Airlines with short and medium-distance flights also may see disruptions. Let's look at short trips of 200 miles or so. These include flights such as Boston to/from New York, New York to/from DC, or Austin to/from Dallas. Airfare each way ranges from $175-$300. But the total cost could be much higher when you include parking and taxi costs. Parking could add $30-$50, depending on the length of time. A taxi on the other end could add $25-$50, although ARS will reduce this cost. In any case, the total cost for the flight, including parking and taxi, ranges from $225 to $400. The cost for an autonomous ride service for longer trips such as this would be similar or lower.

The time required and convenience are also considerations. I know in traveling some of these routes that I would fly for several trips until I got delayed or missed a flight. Then I would drive for a while until I got stuck in traffic or got drowsy driving. With AVs, driving will become more convenient. Depending on the distance to and from the airport, it can take less time. For example, let's assume someone is 45 minutes from the airport and 30 minutes from the airport to the destination. They arrive at the airport an hour before the flight, providing just enough time to park, go through security and board. There are another 30 minutes to get off the plane and get to the taxi. The flight itself is probably an hour and 15 minutes. That's a total of four hours, approximately the same as the travel time for an AV without the hassle of delayed flights. You can also leave on your own schedule without worrying about your flight.

## Other Disruptions

### Loss of Traffic Ticket Revenue

The money from fines for speeding, cruising through stop signs, and lesser infractions like parking violations generate hundreds of millions of dollars for cities, helping to pay for infrastructure and even court budgets. But a city of autonomous, law-abiding cars and trucks, could spell the end of moving violations and traffic tickets.

The total municipal revenue from traffic tickets is difficult to estimate accurately, but it is a considerable amount. One estimate from 2007 is that somewhere between 25 and 50 million traffic tickets are issued each year. Assuming an average ticket cost of $150.00, the total upfront profit from tickets ranges from $3.75 to $7.5 billion. I think the estimates may be a little high, but even at a more conservative estimate of $2.5 billion, a 70%-80% reduction would have a significant impact on many municipalities. New York City, for example, collected $569 million in parking and traffic fines in the 2016 fiscal year.

A decline in revenue from tickets could be more painful in cities and towns that rely more on fines. A 2015 investigation found five Colorado towns relied on traffic fines for at least 30% of their budgets. These fines comprised almost the entire budget of one town. The Nevada Supreme Court complained in 2015 that a decline in traffic tickets was crimping the court's budget. States will be forced to come up with other ways to bridge the gap if and when safe self-driving vehicles become commonplace.

### Oil Industry

Autonomous vehicles will predominately be electric, which as predicted earlier will reduce the need for gas stations. It also will reduce gasoline and oil consumption, but it is difficult to estimate this reduction.

### DUI/OUI

There will be no more DUI/OUI offenses with AVs, eliminating the associated fines, and court costs. The social benefit of eliminating DUI is significant, but the reduction in revenue related to processing these cases will be a disruption for some municipalities. It's not likely to be one that they try to prevent.

### Legal

Many lawyers will lose sources of revenue from traffic offenses and auto crash litigation. The estimated legal costs in the United States are in the billions of dollars. Some laws firms are almost entirely reliant on this type of income and will go out of business.

### Emergency Services

Accidents will be reduced by 90%. Ambulance and other emergency vehicles will be needed less. Hospital emergency rooms will see a reduction in accident patients. One unintended consequence may the reduction in organs available for organ transplant.

## Overall Disruption

The advent of autonomous vehicles will bring amazing benefits, but it will also bring the loss of tens of millions of jobs in the United States alone. Historically, when technology wipes outs jobs, these jobs are replaced by other jobs in new areas. That pattern continues, of course, until it doesn't. Eventually, technology will eliminate so many jobs that they can't be replaced, and society will need to react in some, preferably orderly, way. AVs and autonomous ride services will create new jobs too. Those involved in the development of autonomous technology have already been created. Others will be created in the deployment stage of AVs and ARS. However, these won't offset the sheer number of jobs lost.

Many businesses as we know them today will be diminished or go away. This is part of the natural evolution of industry. Billions of dollars are made with the creation of new industries, while billions are lost in the industries they replace.

In the concluding chapter, I try to estimate the timing of how these disruptions will occur. I'm hopeful that the transitions resulting from autonomous vehicles can be managed in an orderly way, and I encourage governmental leadership to consider constructive solutions.

# Chapter 9
# Government Regulation & Support

Government regulation will be necessary for autonomous vehicles, but my current opinion is that it won't stop or significantly delay the advent of autonomous driving. The significant reductions in death and injuries from accidents, significant decreases in the cost of transportation, and the opportunity to provide mobility to millions who are currently deprived of it, will outweigh the disruptions described in the previous chapter.

As always, there will be some vocal Luddites who are afraid of this new technology, and there will be some reasonably sounding arguments they will make in lobbying for restrictions. They can raise fears that software bugs in cars will kill people, and that cars shouldn't make moral decisions selecting outcomes in different ways to react to an impending accident.

*The significant benefits of AVs outweigh the disruptions.*

There will be concerns expressed about hacking the software in cars, and I expect there will be some best-selling novels along this theme (probably by Steven King). Finally, there will be very well justified arguments about the loss of jobs, as was discussed in the previous chapter. But I don't think that the trade-off of jobs for lives and injuries will prevail. Nevertheless, there is now and will continue to be lobbying to restrict or delay autonomous vehicles through government regulation.

There will also be lobbying for AVs. For example, Waymo launched a public campaign called "Let's Talk Self Driving" with the National Safety Council, Mothers Against Drunk Driving, and the Foundation for Senior Living, among other organizations.

I also believe that there will be governments, particularly at the local level, that will encourage and cultivate the use of autonomous vehicles. It could even become a way for a metropolitan area to compete in attracting new residents and travelers. Eventually, the availability of autonomous rides services and being AV-friendly may become important factors in the rankings of the best places to live.

Since the focus of this book is on autonomous vehicles in the United States, I'll only focus here on the American government. Foreign regulations of AVs are extraordinarily complex and varied. Government regulation in the United States has three levels: national, state, and local. Since these regulations are always in flux, I'll only provide an overall perspective at each level.

## Federal Regulation of AVs

At a national level, regulation focuses on making AVs legal to produce and sell. In 2017, the U.S. House overwhelmingly passed a bill providing the federal government with a framework for developing new rules for driverless cars, which is the first step to allowing AVs vehicles on public roads. At the writing of this book, the Senate is considering similar legislation, and the best guess is that the senate may consider its bill in the fall of 2018. These bills from the House and the Senate signal that the federal government may be ready to tackle the issue and modernize federal regulations to address the new technology.

The House bill is called the Safely Ensuring Lives Future Deployment and Research in Vehicle Evolution, or the SELF-DRIVE Act. The Senate's bill is called the American Vision for Safer Transportation Through Advancement of Revolutionary Technologies, or AV START Act. Cute acronyms, aren't they? Both bills give the National Highway Traffic Safety Administration the authority to regulate the design, construction, and performance of self-driving cars. There are some differences in the directives to the Department of Transportation in the two bills, generally involving timing for new rules and the need for safety evaluation reports.

The most important part of the new regulations is an increase in the number of exemptions car manufacturers have before complying with comprehensive Federal Motor Vehicle Safety Standards. The current number is 2,500 vehicles. Under the House bill, exemptions would begin at 25,000 for the first year and then increase to 100,000 for the third and fourth years. The Senate bill, by comparison, allows the National Highway Traffic Safety Administration to grant up to 15,000 exemptions in the first year. The cap increases to 40,000 self-driving vehicles per manufacturer in the second year and ramps up to 80,000 for subsequent years. To receive an exemption, manufacturers are required to prove to the National Highway Traffic Safety Administration that the car is as safe or safer than cars already on the road.

Both of these include a high number of exemptions that will enable the AV industry to get off to a quick start. There will be many companies launching AVs and the autonomous ride services (ARS) industry, in particular, will benefit. With an exemption of 80,000 to 100,000 vehicles, they can enter many local markets. Remember the displacement ratio for ARS vehicles to individually-owned vehicles is 8-10 to 1.

State and local governments are generally in favor of these bills at the national level but want to see compliance with traffic laws remain at the local level. "With this approach, state and local laws will continue to focus on the operational

154

safety laws regulating motor vehicles and their operators after such vehicles have been constructed and introduced to public roadways," stated the National Governors Association, supported by the National League of Cities and the United States Conference of Mayors.

The federal highway system can also be adjusted to provide an AV-compatible infrastructure, which could include highway lane markings, electronically readable signs, and support for vehicle-to-infrastructure communications. Eventually, national highways may be restructured to include AV-exclusive lanes.

Federal laws may also bar states and cities from implementing "unreasonable restrictions" on the rollout of autonomous vehicles.

## State Regulations

State and local governments regulate registration, licensing, insurance, and safety and emissions inspections. They also regulate and manage their state highway systems.

Some states, 25 currently, have already passed legislation or issued executive orders related to autonomous vehicles, according to the National Conference of State Legislators. Some of these are restrictive. New York, for example, wants to regulate testing, requiring AVs to follow an approved route with a police escort. The state also has a law that requires drivers to have at least one hand on the wheel of any car, though it was suspended until April 2018 to accommodate autonomous vehicle testing.

Some states are promoting the use of AVs. Michigan, California, Arizona, Pennsylvania, and Florida are examples. I believe that eventually, all states will embrace AVs because they won't want to be left behind.

California issues permits for testing autonomous vehicles. As of January 11, 2018, the State of California had issued autonomous vehicle testing permits to 50 companies. Under the testing regulations, manufacturers are required to provide the DMV with a Report of Traffic Accident Involving an Autonomous Vehicle within ten business days of the incident. The California Autonomous Vehicle Testing Regulations also require every manufacturer authorized to test autonomous vehicles on public roads to submit an annual report summarizing the disengagements of the technology during testing.

## Local Regulation and Support

I believe that local regulation and support for AVs will be the primary focal point. Autonomous vehicles may be restricted in some places, but they will benefit from local support in most. Local could be a city, metropolitan area, or county. Since most road systems outside of federal and state highways are managed at local level, they are in a position to promote autonomous vehicles for their area.

ARS is the reason that local support will be favorable. Municipalities that support AVs, particularly ARS, will be priority locations for the initial autonomous ride services deployment. ARS companies will merely bypass those municipalities with restrictions or lack of support. The demand for mobility for seniors and the disabled, the convenience of ARS, and the safety improvements from AVs will be overwhelming. All municipalities will eventually come around to support AVs.

There are some relatively simple things that local governments can do to promote AVs in their jurisdictions:

- They can make sure all roads have painted lines sufficient for the cameras on autonomous vehicles to see.
- There may also be some intersections where the painted lines need to be redone to make it easier for AVs to understand.
- It may be helpful for them to replace four-way stop signs at intersections where drivers need to alternate, because this may be difficult for AVs.
- They can replace or repair any traffic signs that are damaged or not seen by sensors.
- In some cities, it would be helpful to provide defined pick-up and drop-off areas for autonomous cars. There will be a need for an increase of these, as the demand for parking spaces and lots decreases.
- It may be helpful to have reserved temporary parking facilities for autonomous vehicles, similar to defined parking for electric vehicles.
- In some cases, using more left-turn traffic signals could be helpful.
- Eventually, it will be useful to support vehicle-to-infrastructure communications at traffic lights, intersections, etc.

Many of these changes will be most helpful for autonomous ride services, which will be the first primary market for autonomous vehicles. Companies planning to introduce ARS will be willing to review the roads and traffic control systems and make recommendations to a local government. The cost of these changes may not be significant, and given the benefits to the local citizens, many local governments will be willing to make these to entice autonomous ride service companies to enter their markets. It is possible that some local governments may want autonomous ride service companies to fund these improvements. Many ARS companies would be willing to do that in return for an exclusive license in that area for a period, effectively locking out completion.

## Autonomous Driving Test Locations

Early testing of autonomous vehicles has begun, and several cities in the United States are leaders.

### Phoenix

Limited regulations, mild winters, and predictable street grids are turning the Phoenix area into a hotspot for testing self-driving cars. Here are some of the companies doing tests there:

- Intel Corp. is testing Ford Fusion Hybrids on the roads of suburban Chandler.

- General Motors Co. and its Cruise Automation subsidiary are putting a self-driving Chevrolet Bolt EV through its paces in downtown Scottsdale.
- A self-driving Volvo XC90 owned by Uber has been tested in Scottsdale and the surrounding area.
- Waymo plans to have hundreds of real people in the Phoenix area ride along in self-driving Lexus RX450h crossovers and Chrysler Pacifica Hybrid minivans.
- Ford is also testing its autonomous cars in Phoenix in the winter.

Arizona Gov. Doug Ducey in August 2015 signed an executive order on self-driving vehicle testing and pilot programs to help encourage the development of the technology in the state. Automakers and companies don't need special permits or licenses to test in the state. Autonomous vehicles have the same registration requirements as other cars and trucks. Unlike California, Arizona doesn't require companies to report each time that a test driver overrides a car's autonomous mode, and that is important to some companies that want to keep this information confidential.

### Pittsburg

In May of 2016, Uber launched a pilot fleet of autonomous Volvos in Pittsburg. The two companies funded the planned $300 million project. Pittsburgh is home to both the robotics center at Carnegie Mellon University and Uber's engineering Technologies Center. Interestingly, Uber provided Pittsburg with a list of "smart infrastructure upgrades" the city could make:

- Non-exclusive access to specific bus lanes (not busways), which would not be exclusive to the company, but instead would be for any providers of on-demand self-driving services.
- Designation/painting of dedicated lanes in particularly tricky areas and intersections for use by on-demand, self-driving cars.
- Designation/painting of dedicated pick-up and drop-off areas around select spots for on-demand, self-driving cars.
- Improved signaling/signage at certain intersections to optimize the movement of self-driving cars.
- Installation of dedicated short-range communications (DSRC) signals that can be utilized by self-driving cars.
- Installation of bike lanes on select streets, which, in addition to promoting bike use, would create a more accessible/safe environment for self-driving cars.
- Non-exclusive access to municipal parking lots to allow staging of self-driving cars while they are awaiting dispatch.
- Prioritization of snow removal to permit continued service on these routes.

None of these were granted, but they provide some insights into the type of things that an ARS might require for deployment in a municipal area.

Pittsburgh is also an engineering and test center for autonomous vehicles built by Ford, which intends to invest $1 billion over five years in Pittsburg-based Argo AI, specializing in artificial intelligence.

### San Francisco Bay Area

The San Francisco Bay Area is home to much of the technology behind autonomous driving, so it is no surprise that there is much vehicle testing there. Many companies are testing AVs in that area. This includes testing in the Silicon Valley area as well as on the streets of San Francisco. GM is testing in the city.

### Detroit Metropolitan Area

The American Center for Mobility has created a facility at Willow Run for mobility and advanced automotive testing and product development. The center intends to accommodate the broad needs of industry and government with its capabilities to test vehicles, roads, infrastructure and communication systems. The Willow Run site comes with some features and structures including double overpasses, a railroad crossing, and a highway loop to test at sustained highway speeds. Testing can occur during all four seasons, day and night, in sun, rain, ice, and snow. These elements help to create an environment for testing and setting national standards for mobility technologies before vehicles, and other products are deployed. The center will also serve as a development facility that will allow companies to lease space for office and research use, garages and other amenities.

## Summary of Government Regulation and Support

Because of the significant benefits from AVs already discussed, I believe that government regulation at any of the three levels will not impede deployment. There may be some delays to get the regulations completed, but these delays should be minor.

The government, particularly at the local level, will be eager to promote autonomous driving, particularly ARS. In fact, I expect that there will be much competition to secure early ARS operations. The ARS companies will be able to prioritize their deployments based on the support provided by municipal governments.

# Chapter 10
# Stages (Timeline) for AV Adoption

The crucial question is not *will* there be a market for autonomous vehicles, but *when* and *how* will it emerge. In this concluding chapter, I use a stages model to predict the timing of adoption for the major AV markets. Each stage is five years long, and it's important to note that the significant AV markets will emerge differently in each stage.

Categorizing the timing of the adoption of AVs into stages enables a more strategic view. The exact timing of what will happen in each year within a stage is difficult to predict, but the overall perspective provided by stages can be valuable in setting strategy and anticipating the general timing. Before looking at each of the stages of AV adoption, let's consider technology adoption rates in general.

## Technology Adoption Rates

A look at the adoption rates of previous technologies is an excellent place to start, and research by Michael Felton and the Harvard Business Review provides a comprehensive look at these.

It took the original automobile (the horseless carriage) approximately 20

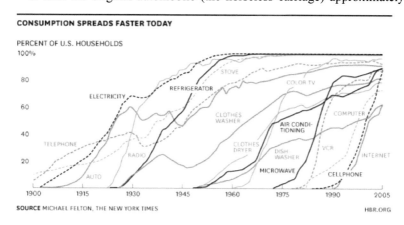

**CONSUMPTION SPREADS FASTER TODAY**

PERCENT OF U.S. HOUSEHOLDS

SOURCE MICHAEL FELTON, THE NEW YORK TIMES

years (from 1910-1930) to get used by 60% of the households. It took another 30 years to get to a penetration of 80% because of wars and setbacks in the economy. Electricity, which required a great deal of infrastructure, took 40 years to get to a penetration of 80% of the households.

Technology adoption has accelerated in more current times. Color TV took about 15 years to get to 80% household adoption, and likewise, cellphones reached 80% household adoption in approximately 15 years.

There are several factors to consider when trying to estimate the adoption rate for autonomous vehicles. Autonomous ride services (ARS) will have a relatively fast adoption rate, for reasons previously discussed. People don't need to make a major purchase; they just need to summon a ride on their cellphone and try it. The cost is low, and the risk is minor.

I expect that ARS will be deployed regionally, so adoption rates will be relative to the number of households using it in a region where it is available.

The adoption rate for individually-owned AVs will be progressive. Vehicles will get increasingly autonomous, so it will be difficult to perceive when they cross into the categorization of sufficiently autonomous. But the adoption rate for some of the features is already fast. Data published by Bloomberg Businessweek shows the rapid adoption of onboard cameras and radar sensors, as is seen in the following chart:

## Cars Smarten Up

■ Annual sales of vehicles with onboard cameras  ■ Annual sales of vehicles with radar sensors

Data: Bloomberg New Energy Finance, MarkLines; graphic by Bloomberg Businessweek

Overall, I expect the adoption rate for ARS to reach 50% by the end of Stage 2 in 2030, meaning that 50% of the households have used some form of autonomous ride service. The adoption rate for individually-owned AVs will take longer since this is a replacement for current automobiles. I believe that it will take approximately 15-20 years until 2035-2040 for 80% of the vehicles to be fully autonomous.

# Stage 0 (2016-2020) – Development and Testing

Some early autonomous vehicle research started before 2016. Google, for example, started its initial research effort in 2008. But by around 2016, many companies began to make serious commitments to developing AVs. For instance, in 2016, Google launched Waymo as its AV company.

During this initial stage, which I refer to as Stage 0, major companies are investing more than $100 billion to develop and test AVs. More than a hundred companies are involved in AV development now in differing ways. At the mid-point of this stage, more than a dozen companies will be actively testing their initial AV prototypes.

Toward the end of this stage, by 2019-2020, we will see the first launch of AVs without drivers, although it will be limited in the number of vehicles and restricted to a few geographic areas. At the conclusion of Stage 0, AVs will be proven to be viable, economical and safe.

# Stage 1 (2021-2025) – Launch of ARS

As discussed previously, autonomous ride services (ARS) will be the first primary market for AVs, and it is the primary focus of Stage 1.

## Stage 1: ARS Market

ARS will be a new market created by some very aggressive companies including Waymo, Uber, Lyft, Apple, GM, Ford and possibly others. There will be a massive "land rush" to secure metropolitan areas for ARS by being the first to get established in that market. I expect that there will be a flurry of activity across the country that will dominate the press and discussion.

To illustrate how fast this market will expand, I have done the following estimates of ARS penetration, growth, and revenue:

|  | 2021 | 2022 | 2023 | 2024 | 2025 |
|---|---|---|---|---|---|
| ARS Vehicles (thousands) | 50 | 150 | 300 | 600 | 1,000 |
| ARS Revenue (millions) | $7,656 | $22,969 | $45,938 | $91,875 | $153,125 |
| ARS passenger miles (millions) | 6,125 | 18,375 | 36,750 | 73,500 | 122,500 |
| Percent of total miles driven | 0.19% | 0.57% | 1.15% | 2.23% | 3.50% |
| ARS miles per person per year | 18 | 54 | 107 | 213 | 353 |
| Number of fleet centers | 125 | 375 | 750 | 1,500 | 2,500 |

At the end of Stage 0, there will be several thousand ARS vehicles in the United States mostly completing pilot programs. By 2021, the first year of Stage 1, I project there will be approximately 50,000 ARS autonomous vehicles in the United States. These vehicles will be deployed in the first 100 or so fleet centers in the early 50 or so metropolitan areas in the United States that embrace ARS. At this point, ARS is still a tiny portion, less than 0.2% of the total miles traveled in the country. Nevertheless, by 2021, the ARS market will already be greater than $7.5 billion.

By the end of Stage 1, I expect that the number of ARS vehicles will increase significantly to an estimated 1 million, possibly more. The ARS market will be greater than $150 billion, and there will be approximately 2,500 ARS fleet centers in the country, located in more than 1,000 municipal areas. Although a very large industry by the end of Stage 1, the penetration of ARS is still relatively low at approximately 3.5% of the estimated 3.5 trillion total miles traveled in the United States, indicating an enormous future potential.

During Stage 1, there will be a flurry of competitive activity as companies try to seize this emerging opportunity. Real partnerships will emerge from the tentative ones created in Stage 0. The structure of the ARS market will develop as provisional strategies evolve into concrete ones. Overall, more than $200 billion will be invested by ARS companies during Stage 1 to establish their initial vehicle infrastructure.

Crucial strategic questions will be answered in Stage 1. Will Apple compete in this market? Who will make the cars for Waymo and how will it promote its ride services? Will Lyft remain an independent company with multiple partners using its open platform or will it be acquired? How will GM, Ford, Mercedes, and other auto manufacturers compete in the ARS market? How will ARS customers show their preferences for different ARS companies? What companies will be perceived as the leaders in ARS? How much will the market valuations of ARS companies jump based on initial success?

*Crucial strategic questions will be answered in Stage 1.*

By the end of Stage 1, ARS will be established as a standard and accepted way of travel. The supply of autonomous ARS vehicles will have a difficult time keeping up with demand.

## Stage 1: AV Retail Market

The retail AV market will begin to grow during Stage 1, but it will be a progressive form of autonomous driving. Individually-owned AVs will progress through Level 3 to Level 4 to Level 5 of the SAE autonomous driving categorization.

Premium-priced vehicles will be the primary beneficiaries of autonomous technology in Stage 1. These buyers will be willing to pay more for autonomous features than will other car buyers. Initially, autonomous capabilities will be features; then, particularly in high-end models, all vehicles will be autonomous.

The level of autonomy will improve annually during this stage with some of the improvements made by software updates. Drivers of these increasingly autonomous vehicles will decide when to engage autonomous capabilities. Some will use them regularly while others may not be as comfortable during this stage. Individually-owned AVs in this stage will probably not be capable of driving everywhere, so drivers will need to take control of the vehicle from time to time. For this reason, individual-owned AVs in this stage will have driver controls and a steering wheel, although the design may be more minimal.

To get some feeling for the transition to autonomous vehicles in independently-owned vehicles, let's use sufficiently-autonomous as the measure. I expect that the number of new car sales will decline as ARS penetration increases. The decline will be gradual in Stage 1, from an estimated 17.5 million sold in the United States (like current sales) to approximately 16 million by 2025. Sufficiently autonomous vehicle sales will increase from less than 0.5% at the start of Stage 1 to an estimated 5% by 2025. At that point, approximately 800,000 sufficiently autonomous vehicles will be sold per year. However, this will still only represent less than 1% of the registered vehicles. Note in my estimates that the number of registered vehicles in the United States is also expected to start to decline in Stage 1.

| | U.S. Individually-Owned AV Estimates | | | | |
|---|---|---|---|---|---|
| | 2021 | 2022 | 2023 | 2024 | 2025 |
| Vehicles Registered (million) | 265 | 263 | 262 | 260 | 250 |
| New Cars Sold (million) | 17.5 | 17.0 | 16.5 | 16.2 | 16.0 |
| Pct. Sufficiently Autonomous | 0.4% | 1.0% | 1.5% | 3.0% | 5.0% |
| Total AVs Sold (thousand) | 70 | 170 | 248 | 486 | 800 |
| Total AVs registered (thousand) | 70 | 240 | 488 | 974 | 1,774 |
| AVs Pct. of Registered | 0.03% | 0.09% | 0.19% | 0.37% | 0.71% |

The number of sufficiently autonomous individually-owned vehicles in Stage 1 depends on the success Tesla has in enabling sufficiently-autonomous features on cars it had previously sold, as it promised to customers who bought them. It also depends on the percentage of customers who enable and use these sufficiently-autonomous features. The estimate of 800,000 sufficiently-autonomous vehicles by 2025 could be higher or lower by several hundred thousand.

## Stage 1: Trucks, Delivery, and Bus Markets

In Stage 1, the market segments for autonomous trucks, delivery vehicles, and buses will begin to advance with initial penetration in each of these segments.

The first penetration of AV trucking will be in the long-haul market. I think that during this stage, AV technology will supplement drivers rather than replace them. Drivers will still be on board but will be able to sleep and rest during significant stretches of driving. Drivers won't be required to take the 8-hour breaks because the truck will be operating autonomously, but they will be there to take control at the beginning and end of the trip as well as handle refueling. The use of autonomous technology combined with drivers will provide enormous benefits with 40% improvement in shipping time and almost the equivalent in truck asset utilization. I estimate that less than 2% of the large trucks used for longer hauls, about 150,000, will be retrofitted or built to do this by the end of Stage 1.

Autonomous delivery will start in Stage 1 but probably won't begin to ramp up significantly until the end of the stage. Autonomous package delivery will be hampered by the final handoff problem of needing to have someone take the package from the delivery truck. There will also be some slowdown in adoption created by labor unions. For example, the Teamsters Union, representing 1.5 million members, is trying to negotiate to get UPS to agree not to use self-driving vehicles for 260,000 UPS employees. I do expect some penetration from an autonomous delivery vehicle following the driver/delivery person as described in Chapter 5.

Autonomous food delivery will start to ramp up early in Stage 1 and will grow significantly by the end of the stage. The benefits of autonomous food delivery are significant, and it doesn't have the problem of taking food from the delivery vehicle that package delivery faces. The initial delay in this stage will be the time required for food delivery companies to design and procure specialized delivery vehicles. Once they get into the procurement and deployment toward the end of Stage 1, I expect the ramp-up will come fast. There will be some lag time early in the stage for customer acceptance.

Autonomous shuttle services will be deployed relatively quickly in Stage 1. They will follow relatively fixed and straightforward routes and offer significant benefits in cost savings and customer convenience. AV manufacturers that can design and manufacture autonomous shuttles in volume will be successful in Stage 1.

Autonomous bus transportation, other than shuttle services, will grow much more slowly in Stage 1. In fact, ARS may displace bus services, so the financial benefits of autonomous bus services will be limited. Many forms of autonomous bus transportation face challenges of autonomously loading and unloading passengers.

### Stage 1: Disruptions

It's helpful to speculate what disruptions described in Chapter 8 will start in Stage 1. Since the primary growth of AVs will be in autonomous ride services (ARS), most of the disruptions will stem from that.

As more people use ARS in this stage, the demand for car sales will begin to decline gradually. I expect this decline will be more in the mid-priced and lower-priced ranges where people primarily use cars for transportation. In particular, families that own more than one vehicle will reduce the number they own and use ARS instead. Other auto segments won't be as affected in Stage 1. The market segment for premium-priced cars will probably not decline as much because reducing the cost of transportation will not be as crucial in this segment. Also, the pick-up truck market will not drop since these are used for more purposes than transportation.

Industries supporting retail car sales will see a gradual decline in Stage 1. These industries include auto dealers, car repair, insurance agents, etc. It won't be destructive for most of them in this stage, but they will see their level of business decline, with profits declining faster. I also expect that the resale value of used cars will begin to drop precipitously in this stage.

## Stage 2 (2026-2030) – Broad AV Acceptance and Disruption

AVs will become fully entrenched as a primary form of transportation in the United States, as well as other countries, during Stage 2. In Stage 2, we will see the magnitude of benefits of AVs discussed in Chapter 3. Annually, tens of thousands of lives will be saved, and hundreds of thousands of people will avoid serious injuries. The cost of transportation will be dramatically lowered. And millions of people previously unable to travel will now have their mobility.

AVs will connect with each other and the infrastructure in Stage 2. The number of AVs will reach a sufficient critical mass that this communication will provide substantial advantages. During this stage, I expect that AVs will get "preferred treatment" such as dedicated high-speed lanes, which will further accelerate the demand for individually-owned AVs to replace non-autonomous cars in addition to increased use of ARS.

ARS will emerge as an enormous market. AVs will dominate the automotive retail market. Driverless delivery will become commonplace. And the disruptions predicted in Chapter 8 will become a center of attention.

### Stage 2: ARS

The ARS market will become one of the largest worldwide markets ever. The number of ARS vehicles in use in the United States alone will increase from 1 million at the end of Stage 1 to 5 million at the end of Stage 2. As can be seen in the chart that follows, the increase will be rapid throughout Stage 2 with the supply of ARS vehicles being the primary constraint.

The combined revenue of ARS companies in the United States will be more than $500 billion by 2028 and more than $750 billion by the end of Stage 2. They will be providing more than 600 billion passenger miles by the end of 2030, but to keep this in perspective, it will only be about 15% of total miles driven in the United States.

| | U.S. ARS Market Estimates | | | | |
|---|---|---|---|---|---|
| | 2026 | 2027 | 2028 | 2029 | 2030 |
| ARS Vehicles (thousands) | 1,800 | 2,600 | 3,500 | 4,300 | 5,000 |
| ARS Revenue (millions) | $275,625 | $398,125 | $535,938 | $658,438 | $765,625 |
| ARS passenger miles (millions) | 220,500 | 318,500 | 428,750 | 526,750 | 612,500 |
| Percent of total miles driven | 6.13% | 8.61% | 11.28% | 13.17% | 15.31% |
| ARS miles per person per year | 630 | 897 | 1,191 | 1,443 | 1,706 |
| Number of fleet centers | 3,600 | 4,333 | 5,833 | 7,167 | 8,333 |

The ARS leaders will emerge in Stage 2 with a few significant competitors dominating the market in each country, and a few companies competing in multiple countries. These large ARS companies will have several hundreds of billions in revenue with substantial profits. Their stock market capitalizations will exceed $1 trillion, making them some of the most successful companies ever.

The total investment to get there will be staggering. I estimate that more than $600 billion will be invested in ARS in Stage 2, including replacement of the 5 million ARS vehicles deployed in Stage 1.

## Stage 2: AV Retail Market

The AV retail market will become the primary focus in Stage 2. More vehicles will be designed and manufactured to be sufficiently autonomous, and the price for these vehicles will come down. Most customers will want the functionality and convenience of sufficiently autonomous vehicles and will trade in their cars for them. However, this is when the value of used, non-autonomous, cars declines. Trade-in values will be much lower. The demand for used "dumb cars" will be very low, and many of these will be shipped offshore to developing countries to salvage some value.

The following table shows the significant changes in the retail vehicle market, including the growth of AV retails sales.

| | U.S. Individually-Owned AV Estimates | | | | |
|---|---|---|---|---|---|
| | 2026 | 2027 | 2028 | 2029 | 2030 |
| Vehicles Registered (million) | 225 | 200 | 180 | 150 | 130 |
| New Cars Sold (million) | 15.0 | 13.0 | 11.0 | 10.0 | 9.5 |
| Pct. Sufficiently Autonomous | 7.0% | 10.0% | 15.0% | 25.0% | 40.0% |
| Total AVs Sold (thousand) | 1,050 | 1,300 | 1,650 | 2,500 | 3,800 |
| Total AVs registered (thousand) | 2,824 | 4,124 | 5,774 | 8,274 | 12,074 |
| AVs Pct. of Registered | 1.25% | 2.06% | 3.21% | 5.52% | 9.29% |

As the penetration of ARS accelerates, the number of new cars sold will decline precipitously to less than 10 million per year by the end of Stage 2. Similarly, the number of vehicles registered in the United States will decrease to approximately 130 million. Each of these represents about half of what the numbers were ten years earlier. Most likely, pick-up trucks and expensive sports cars will make up more of the mix.

During Stage 2, the percentage of sufficiently autonomous vehicles in new vehicles sales will increase from approximately 7% to 40%. But even at the end of the stage, sufficiently autonomous vehicles will account for only 12% of the vehicles on the road.

By the end of this stage, non-autonomous cars will be considered dangerous, and there will be discussions of legislation to prohibit them.

## Stage 2: Truck and Delivery Markets

Stage 2 will see a significant surge autonomous trucking. Most new long-haul class 8 trucks will include the self-driving capabilities previously discussed. An estimated 200,000 - 300,000 new autonomous long-haul trucks will be built annually. While the number of trucks needed for long-haul shipments will decrease because of the 40% productivity improvement, in Stage 2 this decline will be offset by replacement purchases of autonomous trucks. By the end of this stage, approximately 1 million large trucks will be autonomous, approximately 40% of the 2.5 million trucks in service. The total number of trucks needed is estimated to be down from approximately 3.5 million today.

Autonomous food delivery will become the norm in Stage 2. More than 500,000 specialized autonomous delivery vehicles will be used to deliver pizza, fast food, and other meals. Everyone will become used to seeing these cute little colorful vehicles buzzing around town. Acquiring these vehicles will require an investment of approximately $15 billion by the companies providing autonomous delivery. The major pizza chains will lead the way with specially-designed delivery vehicles that will be much smaller than cars today. Independent delivery services will offer an autonomous delivery service for restaurants and others. These services will become so pervasive that individuals and groups will also use these services. Families can have meals delivered to their elderly parents, churches can have meals delivered to shut-ins, and entrepreneurs will create restaurants in their homes with autonomous delivery.

Autonomous truck rental will emerge as a new market segment. People won't need to drive a truck, just load and unload it.

### Stage 2: Disruptions

During Stage 2, the full-scale impact of the AV disruptions will occur. Tens of millions of jobs will be lost in the United States. This loss of jobs will cause protests and lobbying to slow the progress of autonomous vehicles, but it won't stop what is inevitable. The United States, and the world, will need to figure out how to adjust society to fewer jobs.

## Stage 3 (2031-2035) – More Advanced AVs

By this stage, AVs will become the dominant form of transportation. Vehicle-to-vehicle and vehicle-to-infrastructure communications will enable even higher AV performance. Most traffic will be synchronized with vehicles rushing through intersections coordinated by computers.

ARS will become wholly entrenched, accounting for more than 30% of the miles traveled, and generating more than $1 trillion in annual revenue in the United States. Globally ARS will be a $2 trillion market. The cost per mile of an ARS will be well below $1.

*ARS will be fully entrenched, accounting for more than 30% of the miles traveled.*

There will be as many as 7.5 million ARS vehicles in the United States.

By the end of Stage 3, There will be less than 100 million registered vehicles in the United States, and more than half of these will be sufficiently autonomous. Because of the inroads of ARS, new car sales will be less than 8 million per year, and almost all will be sufficiently autonomous.

This will also be the stage where the containment or possible elimination of traditional non-AVs will be discussed. Will they be considered too dangerous by the new AV safety standards? Will they be restricted to limited roads or entirely outlawed?

The trucking industry will fully adopt autonomous driving in Stage 3. More than 60% of the trucks, carrying most of the shipping, will be sufficiently autonomous to do the tasks required. The remaining trucks will do specialized jobs.

The full consequences of the disruption from AVs will come center stage. Governments will begin to discuss public policies to mitigate these disruptions.

## Stage 4 (3036+) – AVs Completely Displace Cars

By Stage 4, AVs will be the norm, and non-autonomous vehicles will go the way of the horse. They will still be driven, but mainly for recreation in specific areas. All AVs will be connected, and traffic will synchronize at high speeds with almost no accidents. People won't need driver's licenses, and a large percentage of the population will never drive a car. ARS will account for more than half of the miles traveled, and the cost per mile will be less than $0.50, possibly much lower.

## Conclusion

It's clear that autonomous vehicles (AVs) will replace cars and trucks as we know them today. We are currently in Stage 0, the development and testing stage. The launch of autonomous vehicles will begin in 2021, the start of what I refer to as Stage 1. The fastest and most profound change in this first stage will be the advent of autonomous ride services (ARS), which will begin to replace individually-owned cars. ARS will become one of the largest industries in the world.

By Stage 2, starting in 2026, autonomous vehicles will become fully accepted, and most new cars will be sufficiently autonomous. Autonomous trucking will be the focus of much investment. The benefits of AVs will be realized, but so will the disruptions. Both will be substantial. Stage 3 will usher in even more advanced AVs, and by its end, the transition to AVs will be complete. In Stage 4, 15-20 years after their introduction, AVS will completely replace dumb cars as we know them today.

At that time, the horseless-carriage, a little more than a century after its introduction, will pass into history, replaced by the driverless autonomous vehicle.

# Index